Lecture Notes in Economics and Mathematical Systems

Managing Editors: M. Beckmann and H. P. Künzi

Economics

127

Environment, Regional Science and Interregional Modeling

Proceedings of the International Conference on Regional Science, Energy and Environment II
Louvain, May 1975

Edited by M. Chatterji and P. Van Rompuy

Springer-Verlag
Berlin · Heidelberg · New York 1976

Editorial Board
H. Albach · A. V. Balakrishnan · M. Beckmann (Managing Editor)
P. Dhrymes · J. Green · W. Hildenbrand · W. Krelle
H. P. Künzi (Managing Editor) · K. Ritter · R. Sato · H. Schelbert
P. Schönfeld

Managing Editors

Prof. Dr. M. Beckmann
Brown University
Providence, RI 02912/USA

Prof. Dr. H. P. Künzi
Universität Zürich
8090 Zürich/Schweiz

Editors
Manas Chatterji
School of Management
State University of New York
Binghamton N. Y. 13901/USA

Paul Van Rompuy
Department of Economics
Centrum voor Economische Studiën
(Katholieke Universiteit Louvain)
Van Evenstraat 2B
3000 Louvain/Belgium

Library of Congress Cataloging in Publication Data

International Conference on Regional Science, Energy,
   and Environment, Louvain, 1975.
   Environment, regional science, and interregional
modelling.

   (Proceedings of the International Conference on
Regional Science, Energy, and Environment, Louvain,
May 1975 ; v. 2) (Economics) (Lecture notes in eco-
nomics and mathematical systems ; 127)
   1. Environmental policy--Mathematical models--
Congresses. 2. Regional economics--Mathematical
models--Congresses. I. Chatterji, Manas, 1937-
II. Rompuy, Paul van. III. Title. IV. Series:
Economics (Berlin, New York) V. Series: Lecture
notes in economics and mathematical systems ; 127.
HT391.I475 1975 vol. 2 [HC79.E5] 309.2'5'08s
                         [301.31]   76-10461

AMS Subject Classifications (1970): 90 A 15, 90 B 05, 90 B 20, 90 B 99, 90 C 50

ISBN 3-540-07693-X Springer-Verlag Berlin · Heidelberg · New York
ISBN 0-387-07693-X Springer-Verlag New York · Heidelberg · Berlin

This work is subject to copyright. All rights are reserved, whether the whole
or part of the material is concerned, specifically those of translation, re-
printing, re-use of illustrations, broadcasting, reproduction by photocopying
machine or similar means, and storage in data banks.
Under § 54 of the German Copyright Law where copies are made for other
than private use, a fee is payable to the publisher, the amount of the fee to
be determined by agreement with the publisher.
© by Springer-Verlag Berlin · Heidelberg 1976
Printed in Germany
Printing and binding: Beltz Offsetdruck, Hemsbach/Bergstr.

CONTENTS

INTRODUCTION	V
CONTRIBUTORS	IX
ENERGY, ENVIRONMENT AND REGIONAL DEVELOPMENT	1
L.H. Klaassen	
SOCIAL COSTS AND ENVIRONMENTAL CAPACITIES	16
H. Onoe	
THE THEOREM OF PUBLIC OVER-EXPENDITURE : THE ENVIRONMENTAL OVERABATEMENT PROBLEM APPLIED TO BELGIAN HIGHWAY TRAFFIC	23
W. Desaeyere	
LIMITATIONS OF REGIONAL AND SECTOR-SPECIFIC ECONOMIC GROWTH BY POLLUTION RESTRICTIONS AND SCARCITY OF RAW MATERIALS - A REGIONALIZED MULTISECTOR MODEL	38
G. Rembold	
INDUSTRIAL INVESTMENT MODEL IN AN UNDERDEVELOPED AREA	60
P. Migliarese and P.C. Palermo	
COST-BENEFIT ANALYSIS AND OPTIMAL CONTROL THEORY FOR ENVIRONMENTAL DECISIONS : A CASE STUDY OF THE DOLLARD ESTUARY	74
P. Nijkamp and C. Verhage	
ENVIRONMENT AND POPULATION OPTIMUM	111
B. Felderer	
SPATIAL EQUILIBRIUM IN THE DISPERSED CITY	132
M. Beckmann	
A RATIONALE FOR AN URBAN SYSTEMS MODEL	142
C. G. Turner	
MOTIVATION IN SUBURBAN MIGRATIONS RELATED TO ENVIRONMENTAL STANDARDS. AN ANALYSIS OF THE ANTWERP REGIONAL MIGRATIONS	158
M. Van Naelten	
ON MULTI-REGIONAL MODELLING	168
J.H. Paelinck, A. Van Delft, L. Hordijk and A.P. Masterbroek	
INTERREGIONAL ATTRACTION THEORY : A GENERALISATION OF ATTRACTION THEORY THROUGH INTERREGIONAL INPUT-OUTPUT	184
D. Van Wijnsberghe	
NEW INTERNATIONAL ECONOMIC ORDER AND RESOURCE ALLOCATION IN DEVELOPING COUNTRIES	198
P.N. Mathur	

INTRODUCTION

This second volume of proceedings of the International Conference on Regional Science, Energy and Environment (Louvain, May 1975) contains papers related to general and partial equilibrium models of regional and urban development, in which natural and human resources play a dominant role.

It need not be stressed that environmental factors and resource management have, to some extent, been neglected in postwar economic research. Unfortunately, a world-wide energy crisis or more local environmental disruptions were necessary to draw the economist's attention on the increasing imbalance between man and environment.

The topics treated in this volume reflect the shift in economic research which has taken place since the early seventies. They can be classified roughly into 4 fields. The first field deals with a welfare approach to environmental deterioration. The second area covers models of resource allocation that contain environmental constraints. The third class of problems focuses on the relationship between environment and urban development. Finally, some methodological papers are included that explore new areas in regional and interregional model building.

Klaassen opens this volume with a paper on the impact of rising energy prices on the structure of regional development and environment. He analyses the change in size of all potentials and the consequent decrease in the volume of traffic. Besides these short-run influences, a reallocation of households and firms may be expected in the long-run.

In a free market system, where firms maximise profits, a substantial part of social costs, that reflect the depletion of environmental capacities, will be shifted towards consumers. *Onoe* shows that an optimal allocation of costs will be realised if national economic policy intervenes and applies the P.P.P.-principle (polluters pay principle).

On the other hand, *Desaeyere* points to the misallocation that results from government overexpenditure on public goods, which contrasts sharply with Galbraith's statement on public underexpenditure. The overexpenditure theorem is proved assuming (a) that government is reluctant from imposing a social cost pricing system and (b) that government policies tend to eliminate pollution completely.

An operational general equilibrium model of resource allocation with environmental constraints has been elaborated by *Rembold*. Based on input-output theory, the model contains blocks on production, pollution, final demand and trade. A dynamic programming formulation of the complete model is presented as well.

Migliarese and Palermo have constructed a planning model that yields a solution to a constrained investment choice problem and to a constrained industrial location model.

A link between traditional cost-benefit analysis and optimal control theory has been established by *Nijkamp and Verhage* in an attempt to solve a constrained public investment choice problem. The model has been successfully applied to a public waterworks project in the Netherlands. The determination of the optimal size of population can also be looked upon as a long run resource allocation problem with environmental constraints.

Felderer analyses this problem by means of a neoclassical production function combined with a function for negative externalities. Characteristics of the static and dynamic solution are discussed as well as the impact of various types of technical progress.

Beckmann introduces interaction among households in an urban growth model. The impact of the interaction factor, which is supposed to be a variable in the household utility function, has been derived from equilibrium growth conditions for a linear city.

Turner has elaborated an operational urban systems model which allows for a variety of policy measures related to transportation and infrastructure investment, energy and land use. An interesting contribution shows up in the use of efficiency and equity criteria for alternative policy measures.

As noted above, environmental preferences have for a long time been neglected in spatial models. Their impact on the location of households and firms appears important on apriori grounds.
Van Naelten presents the results of a large scale enquiry on environmental preferences of migrants in the Antwerp region, in which a significant environmental influence on the motivation of migrants shows up.

The introduction of a spatial dimension in theoretical model building opens a new field of interesting research. *Paelinck, Hordijk, Mastenbroek and Van Delft* have constructed a framework in which spatial characteristics and economic behaviour have been integrated. The authors show that the use of band matrices is a promising tool of analysis in a discrete spatial version. A continuous version of time-space development is analysed as well.

Van Wijnsberghe generalizes attraction theory in an interregional input-output model, assuming that regional output depends on output, final demand and communication costs of all other regions. This approach opens an interesting perspective for the structural analysis of regional development problems.

Finally, *Mathur* uses a comprehensive interregional programming model in order to derive an economic order or "rules of the game" which would contribute to an optimal interregional resource allocation. The analysis of a submodel for a common market of a group of developing countries reveals that the market price system does not allocate resources efficiently.

Paul Van Rompuy,
Louvain, February 1976.

CONTRIBUTORS

M. BECKMANN	Technische Universität München W. Germany, and Brown University, U.S.A.
W. DESAEYERE	Economische Hogeschool Limburg, Diepenbeek, Belgium
B. FELDERER	University of Köln Köln, W. Germany
L. HORDIJK	Netherlands Economic Institute Rotterdam, The Netherlands
L.H. KLAASSEN	Netherlands Economic Institute and Erasmus University Rotterdam, The Netherlands
P. MIGLIARESE	Politecnico di Milano Milan, Italy
A.P. MASTENBROEK	Netherlands Economic Institute Rotterdam, The Netherlands
P.N. MATHUR	University College of Wales, Aberystwyth, England
P. NIJKAMP	Free University Amsterdam, The Netherlands
H. ONOE	Kyoto University Kyoto, Japan
J.H.P. PAELINCK	Netherlands Economic Institute and Erasmus University Rotterdam, The Netherlands
P.C. PALERMO	Politecnico di Milano Milan, Italy
G. REMBOLD	University of Karlsruhe Karlsruhe, W. Germany
C. G. TURNER	Nathaniel Lichfield and Associates London, England
A. VAN DELFT	Netherlands Economic Institute Rotterdam, The Netherlands
M. VAN NAELTEN	Katholieke Universiteit Leuven Louvain, Belgium
P. VAN ROMPUY	Katholieke Universiteit Leuven Louvain, Belgium
D. VAN WIJNSBERGHE	Regional Economic Council for Brabant Brussels, Belgium
C. VERHAGE	Free University Amsterdam, The Netherlands

ENERGY, ENVIRONMENT AND REGIONAL DEVELOPMENT

L.H. Klaassen
Netherlands Economic
Institute and Erasmus University
Rotterdam - The Netherlands

I. Introduction

Recent developments have made us aware of a number of factors that about a decade ago hardly played a role in considerations about regional development. These factors are related on the one hand to the scarcity of raw materials for energy production and on the other hand to the deterioration of nature as a consequence of human activities.

Discussions about both subjects have brought forward several interesting aspects but have not yet led to a definite conclusion as to what the future will look like. Some argue that energy stocks will last only for a few more decades to come if nothing is done to stop exponential growth of industrial production. Others argue that huge areas of the world, such as the oceans, the Amazon area and the interior of Africa and Siberia have been searched only very superficially while it seems very likely that precisely these regions are among the most promising as far as potentials for energy production are concerned.

About nature dim views have been expressed by some who believe that the present trend eventually will lead to the complete degeneration of nature and the ultimate self-destruction of the human race. Others have expressed as their view that the enormous productive capacities of the world could be put to use for producing nature, cleaning the rivers and the air and dealing successfully with the increasing volumes of solid waste. It has also been argued that even the present knowledge of ecological processes would already enable man to create new ecological systems and new landscapes where needed.

Pessimistic views and optimistic views go hand in hand and it seems still early, with our present defective knowledge, particularly about energy reserves and the technological possibility of producing materials for energy production from hitherto unknown other raw materials, to judge which direction future events will take. Yet it may be safely assumed that, whatever the future course of events, we may be certain that energy will become more expensive, and production costs of goods and services will also rise as a result of the environmental measures taken by the governments.

Some, of course, take another view about energy and judge it desirable for governments to introduce a system of rationing rather than permit a rise in price of energy. It seems that the Arab countries have at least partial solved this dilemma for us. There is a definite rise in price and a greater one to be expected in the future. The consequences for the balance of payments for a large number of both Western and developing countries are not yet clear, however. It might be that the elasticity of demand for energy raw materials is, at least in the short run, so small, that even in the face of much higher prices rationing might be the only possible way out, that is, if governments should prefer this cumbersome measure to a more simple system of taxation on energy consumption equilibrating demand with permissible supply. An advantage of the latter over the first system is, of course, that not only does it limit demand in a rather more simple way, but it stimulates research for new processes as well as the search for new finding places.

In the following we will assume that the major effect of the increasing awareness that conventional raw material stocks are limited will be a (considerable) rise in price of all sorts of energy, and consequently a rise in price of all other products as far as they contain energy. It may be remarked here that the latter rise might be also considerable in spite of the relatively small energy content in most products. Although the energy content per sector is indeed relatively small for most sectors it appears[1] that the cumulative energy content of some sectors is relatively large. For the Netherlands the accumulated energy content of the primary metal sector is at least three times the energy content of the sector itself; for the animal-food industry this figure is five, for engineering industries four, and for the transportvehicles industries about 3 1/2.

It is not clear how governments will deal with these price rises. So far many have pursued a rather peculiar policy in this respect, compensating workers for the full rate of inflation, thus including the

---
[1] Compare "Energy Conservation", Ways and Means, KIVI, Den Haag, 1974, p. 144.

part of "inflation" due to the rise of energy prices. Obviously this policy, which results from the erroneous use of the consumer-price index as a simple measure of inflation, enables the worker to maintain his consumption level and thus also his direct and indirect energy-consumption level. This will reinforce considerably the pressure on the balance of payments, which eventually might lead to devaluation of the currency, with all the consequences for the domestic-consumer price level as well as for the energy prices charged by production countries.

It is obvious that this policy, which understandably is supported by the labour unions, will lead us nowhere. Price increases resulting from real scarcity cannot be compensated. The only result of attempts to do so is a further increase in prices and even an inevitable continued real inflation if the latter is defined as price rises due to an ex ante excess demand.

There is not yet much known about the part of the overall increase in the consumer-price index that is due to the rise of energy prices, but given the high accumulated energy quota of a number of important sectors it may be expected that the influence of increased energy prices on many retail prices is certainly not negligible, particularly not if the additional effects of the compensation policy are taken into account.

In view of the foregoing considerations we formulate our problem now as : What will be the influence of rising and uncompensated prices on the regional structure of the economy ? The assumption that prices will remain uncompensated is made as a long-run assumption, since real scarcity prohibits compensation in the long-run. An additional question to be considered is what will be the influence of these price increases on environmental conditions.

II. The concept of a potential in regional economics

A model can hardly be called a regional model if "distance" in whatever form does not play an essential role in it. The difference between general economics and spatial and regional economics is precisely that the distance factor is lacking in the former and plays an important role in the latter. It seems that the concept of a potential deals elegantly with this aspect in regional economics because it introduces in a systematic way the interrelation between regions.

Let us take a simple example to start with.

Write

$$r_{ij} = \frac{x_j e^{-\alpha C_{ij}}}{\sum_j e^{-\alpha C_{ij}}} r_i^* \qquad (1)$$

in which r_{ij} is the number of people seeking recreation in j and living in i. $r_i^* = \sum_j r_{ij}$ is the total number of people recreating in any j and living in i, C_{ij} represents the generalized costs of transportation from i to j as a weighted sum of money costs, time and efforts. x_j represents the attractiveness of j as a recreation area.

The assumption (1) implies that the total attraction exerted by a recreation area in j on inhabitants of i is determined by the attractiveness of the area itself (x_j) deflated by distance through the e-function. The fraction of the total number of people from i seeking recreation in j is then determined by the ratio of the attraction of j (corrected for distance) and the total attraction of all areas (including j).

We further assume

$$r_i^* = \alpha_o p_i (\pi_i)^{\beta_1} y_i^{\beta_2} \qquad (2)$$

in which p_i = population in i

$\pi_i = \sum_j e^{-\alpha C_{ij}}$

y_i = per capita income in i.

In this function π_i stands for the <u>recreation potential</u> of i. Obviously this potential, which (according to (2)) is one of the main explanatory factors of the total number of recreants, depends on the cost of transportation from i to all j-regions, as well as on all x_j.

The first conclusion we may draw, in so far as we are willing to accept the second hypothesis, is that an increase in all C's - which works out similarly as a corresponding increase in α (the "resistance" against moving) - is that the potential will decrease and thus the total number of recreation-seeking people will decrease.

The second effect may be derived from the first hypothesis.

Write

$$r_{ik} = \frac{x_k e^{-\alpha d_{ik}}}{\sum_k x_k e^{-\alpha d_{ik}}} \qquad (3)$$

then it follows from (2) and (3)

$$\frac{r_{ij}}{r_{ik}} = \frac{x_j}{x_k} e^{-\alpha(C_{ij}-C_{ik})} \qquad (4)$$

This equation shows that a proportional increase in both C_{ij} and C_{ik} increases the difference $C_{ij}-C_{ik}$ and consequently results in a relative decrease of r_{ij}. In other words, the increase in general transportation costs creates a preference for recreation areas closer by. This will result in a decrease in the average distance covered.

The final result is, then, that people will recreate to a lesser extent than before the rise in transportation costs, and cover an average distance shorter than before. Recreation activities will shrink in two ways, in the number of people and in average distance.

Our next step is to define the quality of life in a given region as a weighted product (or sum) of all potentials that are important for the quality of life, such as recreation, housing, environment, education, medical care, job opportunities etc. All these elements can be expressed in potentials equally well as recreation. If educational possibilities are taken as an example, it seems justified to take into account not only the region in which people actually live, but also neighbouring regions, to a degree corresponding with their accessibility from the own region.

If we define quality of life in this way we must come to the conclusion that increased transportation costs influence the value of all potentials and consequently have on the one hand an adverse effect on the quality of life as a result of the decrease in the value of a number of potentials but on the other hand a favourable effect via the decrease in the volume of traffic. If less people move and move a smaller distance if they move, the result will be a decrease in the volume of traffic which, of course, should be judged favourable from a societal point of view.

Less traffic means fewer accidents, less air pollution and less space needed for infrastructure.

Lower potentials mean less recreation, less education, fewer job opportunities, lower environmental quality etc. Which of the two is more important is difficult to judge. The only thing we can say is that if the first factor is more important than the second, then the previous situation has been clearly sub-optimal, which means that taxation on transportation has been too low.

We will not try here to go into the problem of the weights to be attached to different elements in the welfare function. We will confine ourselves to the consequences of increased transportation costs, to be linked later to increased energy prices.

We may now turn to the other end of the scale, the location of production households. The first part of this section was in fact dedicated to the location of households, which supposedly was determined on the basis of the weighted product of all potentials, including job opportunities. It pointed out the factors determining the location of consumption households. Now we will proceed with the location of production households.

In order to keep the argument simple we will start with a sector that is completely demand-oriented, which means that its location is completely geared towards the location of effective demand.

We assume, similarly to what we did in the first part of this section,

$$D_i^p = \sum_j D_j \frac{e^{-\rho C_{ij}}}{\phi_j} \qquad (5)$$

in which D_i^p = demand potential of region i
D_j = effective demand in j
ϕ_j = accessibility index for region j (to be defined later).

(5) represents the weighted sum of demand in all regions j, deflated by communication costs. This concept will require further

attention in a later section. ϕ_j represents the reciprocal of the accessibility of region j from all regions. The easier j is accessible from all directions, the more difficult it will be for a producer in i to penetrate into this market. The value of ϕ_j can be found as follows.

$$\sum_i D_i^p = \sum_j D_j = D \qquad (6)$$

The sum of all potential demands has been realistically defined here as the sum of all effective demands.

Consequently, we may write

$$\begin{aligned} D &= \sum_i \sum_j \frac{D_j}{\phi_j} e^{-\rho C_{ij}} \\ &= \sum_j \frac{D_j}{\phi_j} \sum_i e^{-\rho C_{ij}} \end{aligned} \qquad (7)$$

The obvious solution for ϕ_j is

$$\phi_j = \sum_i e^{-\rho C_{ij}} \qquad (8)$$

The maximum value of ϕ_j is reached for all $C_{ij} = 0$. Then $\phi_j = n$ (number of regions). Its minimum value is reached for all $C_{ij} = \infty$. Then $\phi_j = 0$.

The <u>accessibility coefficient</u> α ($0 \leq \alpha \leq 1$) may now be written as

$$\alpha_j = \frac{\phi_j}{n} = \frac{\sum_i e^{-\rho C_{ij}}}{n} \qquad (9)$$

and (5) may be rewritten as

$$D_i^p = \sum_j D_j \frac{e^{-\rho C_{ij}}}{\sum_i e^{-\rho C_{ij}}} \qquad (10)$$

in which potential demand is defined as a realistic estimate of sales in all regions j of the industry located in i. Since ρ is sector-specific, this equation obviously only holds for a well-defined sector.

Suppose now that C_{ij} increases. The first effect then is that the position of the i-producers on the j-market weakens.

We may easily derive that

$$\frac{\partial D_i^p}{\partial C_{ij}} = -\rho D_{ij}^p \{1 - \frac{D_{ij}^p}{D_j}\} \qquad (11)$$

in which D_{ij}^p represents the sales of i on the j-market. We thus find that apart from ρ the decrease of D_i^p is greater, the larger D_{ij}^p was and the better the relative position of others on the j-market was. In other words, the influence of an increase in generalized transportation costs is great if

a) the competitive position of i on the j-market was weak,
b) the sales in j were large in an absolute sense.

However, since we may assume that in general $C_{ij} = C_{ji}$, there is a second influence of an increase in C_{ij}, viz. that the competitive position of j on the i-market weakens as the transportation costs C_{ji} will also increase.

We find

$$\frac{\partial D_i^p}{\partial C_{ji}} = \rho D_{ii}^p \cdot \frac{D_{ji}^p}{D_i} \qquad (12)$$

in other words, there is a positive effect on the position of the i-producers which is greater the larger the own sales in i were and the stronger the position of j in i was.

The total result equals the algebraical sum of the two effects

$$\frac{\partial D_i^p}{\partial C_{ij}} + \frac{\partial D_i^p}{\partial C_{ji}} = \frac{-\rho D_j e^{-\rho C_{ij}}}{\Sigma_i} \cdot \frac{\Sigma_i e^{-\rho C_{ij}} - e^{-\rho C_{ij}}}{\Sigma_i} + \frac{\rho D_i e^{-\rho C_{ij}}}{\Sigma_j} \cdot \frac{1}{\Sigma_j} =$$

$$= -\rho e^{-\rho C_{ij}} \{ \frac{D_j}{\Sigma_i} \cdot \frac{\Sigma_i e^{-\rho C_{ij}} - e^{-\rho C_{ij}}}{\Sigma_i} - \frac{D_i}{\Sigma_j} \cdot \frac{1}{\Sigma_j} \} \qquad (13)$$

where
Σ_i stands for $\Sigma_i e^{-\rho C_{ij}}$
and Σ_j stands for $\Sigma_j e^{-\rho C_{ji}}$

This expression is negative if

$$\frac{D_j}{D_i}\left(\frac{\Sigma_j}{\Sigma_i}\right)^2 > \frac{\Sigma_i - 1}{\Sigma_i} = 1 - \varepsilon \tag{14}$$

in which ε is a number relatively small compared to one.

This expression obviously can be approximated by

$$\frac{D_j}{D_i}\left(\frac{\Sigma_j}{\Sigma_i}\right)^2 = \frac{D_j}{D_i}\left(\frac{\alpha_j}{\alpha_i}\right)^2 > 1 \tag{15}$$

This implies that the effect will be negative if the j-market exceeds the i-market in size and that to a greater extent, the more accessible the j-market is.

The main conclusion that can be drawn from this exercise is that smaller markets will lose more than larger markets when the level of transportation costs increases, in other words, regional policy directed, as it generally is, towards the development of smaller markets will become more difficult. With the rise in transportation costs the sizes of the relevant regions will decrease and the regional economies will move to a certain extent in the direction of more autarky with less interregional trade and a stronger concentration on the own market.

Evidently, a similar reasoning can be set up for the inputs needed for production. As these can also be expressed as potentials, each region will suffer from a general increase in transportation costs lowering the accessibility of inputs. However, here too, larger, diversified markets will suffer considerably less than smaller markets with a less diversified structure and higher dependence on importation of inputs needed for production.

Both trends show a heavier dependence of local production on local markets and inputs locally available as a result of a general decrease in the values of potentials. It appears that the effects on production households are similar to those on consumer households. The same holds for the labour market. Not only will rising transportation costs decrease the job opportunities for workers, they will just as well influence negatively the availability of workers of the required skill-levels for production households. All potentials will shrink, with both favourable and unfavourable effects. Whether the balance will ultimately be positive or negative will be considered in a later section.

III. Energy prices and communication costs

As already indicated, transportation costs for households should be generally seen as so-called generalized transportation costs. They represent the weighted sum of money costs, time and effort, and the risks of bridging distance.

In order to simplify the argument, we will assume that generalized transportation costs consist of three elements, viz. money costs of transportation, which are assumed to be proportional to distance, time costs, and risk factors, both also proportional to distance.

We then find

$$C_{ij} = (m + r + \frac{\lambda}{v} y) d_{ij} \qquad (16)$$

in which m = money costs per unit of distance
r = risk   "   "   "   "
v = average speed
λy = value of the time unit

Now an essential part of m is energy costs. So, we may write

$$\frac{e_{ij}}{C_{ij}} = \frac{e}{m + r + \frac{\lambda}{v} y} \qquad (17)$$

in which e_{ij} = total energy costs of the journey in j
and   e   = energy costs per unit of distance.

In the foregoing we have considered the consequences of a rise in transportation costs. Here it becomes clear that such a rise may be the result of

a) a rise in energy prices (e)
b) a decrease in risk   (r)
c) an increase in speed   (v)

d) a decrease in income   (y)

We found earlier that a rise in energy costs will lead to a decrease in traffic volume owing to the resulting increased costs of bridging distances. This implies that to the extent that this is true, all above mentioned effects will occur simultaneously, since lower traffic volumes will lead to smaller risks and increases in speed, while a lower level of income results from the exogeneously determined heavy increase in energy prices.

It appears that rising energy prices will result in higher transportation costs together with a heavy increase of the relative share of energy prices in these higher costs.

The situation for production households is somewhat but not much different. The costs of communication are here in many instances more important than actual transportation costs. In fact there are three elements that should be taken into account in the demand- and supply potentials, viz.

a) transportation costs of goods
b) costs of face-to-face contacts
c) costs of telecommunication.

More than consumption households, production households will be able to substitute face-to-face contacts by telecommunication, although it cannot be denied that the quality of this kind of communication is usually inferior to that of face-to-face contacts. The impact of the rise in energy prices will be highest in sectors in which transportation of goods is relative important. This will in general be basis industries. Telecommunication is supposedly most important in the business-service sector, which will be affected least, at least directly, by the increase in energy prices.

The foregoing arguments may show that the rise in energy prices may have considerable influence through the effect they have on transportation costs. Of course, this will be not the only effect. Energy raw materials are also used in other activities than transportation, notably for equipment units and as raw materials in chemical production. This means that the price rise will spread over all sectors via the accumulation of energy-price effects as well as via the accumulation of transportation-price effects.

This development will work out differently for different sectors of the economy and will consequently have different effects upon different regions.

It is, however, the object of this paper to concentrate on the general effects in a spatial context, rather than on the specific influences in specific regions, although the latter no doubt deserve extensive attention. With respect to the specific spatial effects we will now try to draw some general conclusions from the foregoing argument.

IV. Some general conclusions

Schematically, the argument developed so far with regard to consumers' households can be presented as follows.

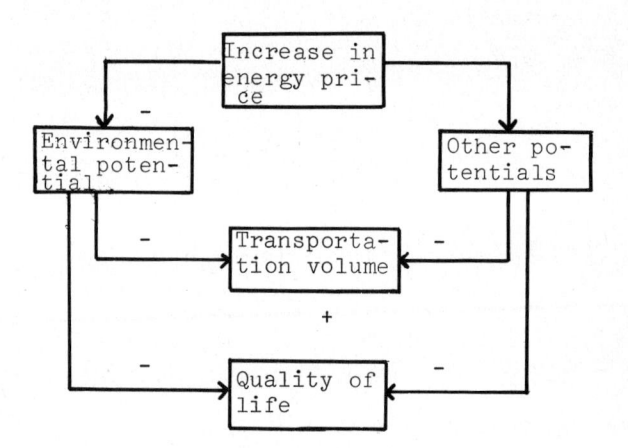

```
                    ┌─────────────────┐
            ┌──────▶│ Increase in     │──────┐
            │       │ energy price    │      │
            │       └─────────────────┘      │
            │ −                              │ ▼
    ┌───────────────┐                ┌───────────────┐
    │ Environmen-   │                │ Other po-     │
    │ tal poten-    │                │ tentials      │
    │ tial          │                │               │
    └───────────────┘                └───────────────┘
            │ −     ┌─────────────────┐   −  │
            │   ───▶│ Transporta-     │◀──   │
            │       │ tion volume     │      │
            │       └─────────────────┘      │
            │               +                │
            │ −     ┌─────────────────┐   −  │
            └──────▶│ Quality of      │◀─────┘
                    │ life            │
                    └─────────────────┘
```

The increase in energy prices will lead, through the effects shown before, to a general decrease in the value of potentials. Although these potentials could be taken together, it seems worth while to mention the environmental potential separately to indicate that the effect of the energy price on this potential is twofold.

The decrease in potentials will eventually lead to a decrease in the volume of transportation. The decrease of the potentials itself can be considered as a negative contribution to the quality of life, the decrease in the volume of transportation, however, as a positive contribution.

For production households the situation is as follows.

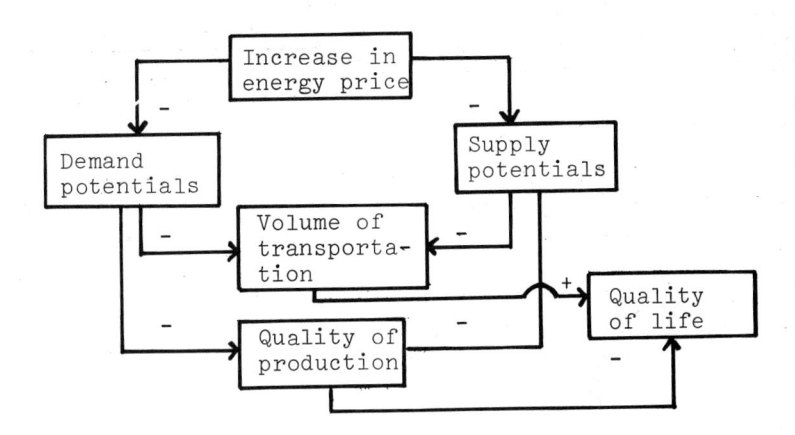

```
                    ┌─────────────┐
          ┌────────→│ Increase in │────────┐
          │      -  │ energy price│     -  │
          ↓         └─────────────┘        ↓
   ┌──────────┐                     ┌──────────┐
   │  Demand  │                     │  Supply  │
   │potentials│                     │potentials│
   └──────────┘                     └──────────┘
        │     ┌──────────────┐     │
        │  -  │  Volume of   │  -  │
        └────→│ transporta-  │←────┘
              │    tion      │
              └──────────────┘           ┌──────────┐
                     │              +    │ Quality  │
        ┌──────────────┐  ─────────────→ │ of life  │
        │  Quality of  │    -            └──────────┘
        │  production  │───────────────────→  -
        └──────────────┘
```

The increase in the energy prices causes a decrease in demand potentials as well as in supply potentials as a result of the decreased accessibility of markets for outputs as well as for inputs. Both decrease lead eventually to a decreases in the volume of transportation which, as it did in the former scheme, favourably influences the quality of life.

The quality of production in general, however, will decrease. It will suffer from the diminished accessibility of both markets and inputs as well as from the more limited availability of qualified workers as a result of the decreased commuting distances implying smaller areas from which workers can be attracted. This influences negatively the quality of life, since the quality of goods and services produced is an element of the quality of life.

Taking both groups of effects together we find

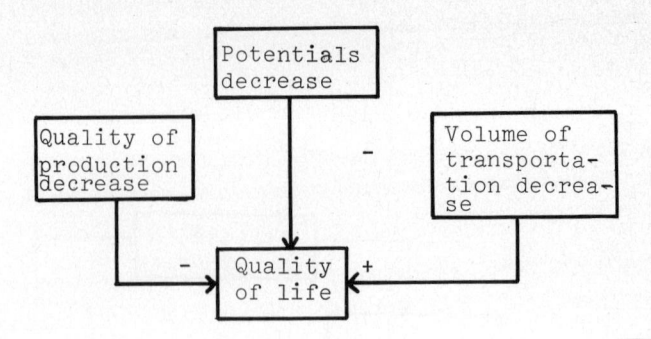

```
          ┌──────────────┐
          │ Potentials   │
          │ decrease     │
          └──────┬───────┘
┌──────────────┐ │        ┌──────────────┐
│ Quality of   │ │        │ Volume of    │
│ production   │ │   −    │ transporta-  │
│ decrease     │ │        │ tion decrea- │
└──────┬───────┘ │        │ se           │
       │         │        └──────┬───────┘
       │    −    ▼    +          │
       └────▶ Quality  ◀─────────┘
              of life
```

This diagram shows that, in fact, the only advantage in spatial development is the decrease in the volume of traffic. All other effects are negative.

As far as we may assume that the energy prices in the new situation represent better than before the real scarcity of energy raw materials we should of course accept the new situation as being better than the original one. Many of us will certainly also welcome the decrease in the volume of traffic as a contribution to the quality of life, even taking into consideration the sacrifices to be made for it in other fields.

We should not forget, however, that so far we have only looked at the changes in the values of the potentials and have not yet considered the impact of these changes on the location of households of both consumers and producers. Under the process of suburbanization workers might have chosen residences rather far from their working place since transportation costs being relatively unimportant in their considerations about their residence. If these cost increase considerably, their opinion might change and they might be inclined to locate closer to their work or to look for another job closer by. Firms might find themselves confronted with heavily increasing expenses for the transportation of their inputs and their final products and might consequently consider a change in location, particularly if they are located in relatively isolated areas, where in most cases they are already leading a marginal existence anyway.

Although these effects are no doubt long-run effects, they should not be taken lightly. It seems likely that high energy prices are here to stay, so that the long-run effects will no doubt materialize. We shall do well, therefore, to study them carefully in order to be able to anticipate the most damaging of them.

Looking at what has been said so far it appears that a great deal of study lies ahead of us. The intention of this paper was only to indicate some of the effects that may be expected in the near or further future. Since these effects are of basic importance for the spatial distribution of our well-being as well as for its overall level, spatial and regional economics are faced with an important task. The size of it should not discourage us from trying to fulfil it to the best of our knowledge.

SOCIAL COSTS AND ENVIRONMENTAL CAPACITIES

Hisao Onoe
Kyoto University
Kyoto - Japan

I. Introduction

When contemporary economists tackle environmental problems, they usually employ the concept of "social costs". But the definition of "social costs" has not yet been firmly established and it sometimes becomes the subject of discussion among economists. This situation arises partly from the fact that this term of economics is closely related to the concept of "environmental capacity" which belongs to the terminology of ecology rather than that of economics. In order to analyse the concept of "social costs" it is essential to examine its relation to the concept of "environmental capacity". Environmental capacity can be defined as the capacity of nature to reduce the quantity of pollutant, such as sulphurous acid gas and nitrogen dioxide. In order to maintain a certain level of environmental capacity, human beings usually employ two methods : restoration or prevention.

II. A Graphical Analysis

To the extent that human beings behave in a rational way, they will try to minimise costs of restoration and prevention in the maintenance or achievement of certain levels of environmental capacity. The relation between environmental capacity and these costs can be expressed by Figure 1, where the isoquant E_0 shows the different strategies viz., restoration and prevention, for a given level of environmental capacity.
Various levels of environmental capacity may be measured by the use of indicators, such as the degree of comfort, sanity and health.

The isoquants slope downwards to the right, because either of these techniques can be substituted for the other. (Here divisibility is assumed). They are convex to the origin, due to the fact that inputs of the two techniques are not perfect substitutes. The least-cost combination of prevention and restoration is found at point A_0 where the isocost K_0 is just tangent to the isoquants of environmental capacity E_0. (Prices of prevention and restoration being constant).

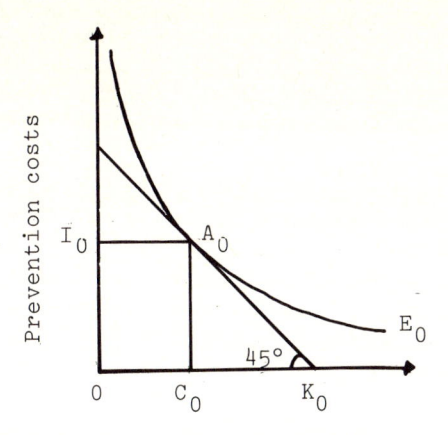

```
                    ↑
                    |
        s           |
        t           |
        s           |
        o           |
        c           |\
                    | \
        n      I₀   |--\------A₀
        o           |   \     |
        i           |    \    |
        t           |     \   |
        n           |      \  |
        e           |       \ |
        v           |        \|
        e           |         \
        r           |         |\         E₀
        P           |         | _____
                    |         |  45°\
                    |_____|_____→
                    0        C₀      K₀

                      Restoration costs

                         Figure 1
```

Rational policy-makers of a communal society will pursue point A_0 and employ techniques of prevention and restoration which cost I_0 and C_0 respectively. In such a society there could be no important reason why the members of the society should be reluctant to achieve the best possible environmental capacity. Thus an environmental capacity which embodies this aim will be satisfactory from the point of view of the existing levels of science and technology. But such a society rarely exists. Japan e.g. is far removed from such a society and is famous for her rapid economic growth and severe environmental disruption.

In a free market society where profit-making enterprises try to minimize their own costs and lighten the burden of the costs of prevention and restoration, realized levels of environmental capacity are likely to be lower than adequate.

Usually, levels of environmental capacity are fixed by the balance of power between polluters and sufferers. Even central or local governments who play the role of arbitrator are apt to be influenced by the state of this balance of power. Hence, these levels will change over time and vary from country to country.

Fig. 2 shows that in a society where private profit-making enterprises prevail, the level of environmental capacity is represented by curve E_1, which is situated nearer to the origin than E_0. The firms operating along E_1 will employ a least-cost combination of remedies

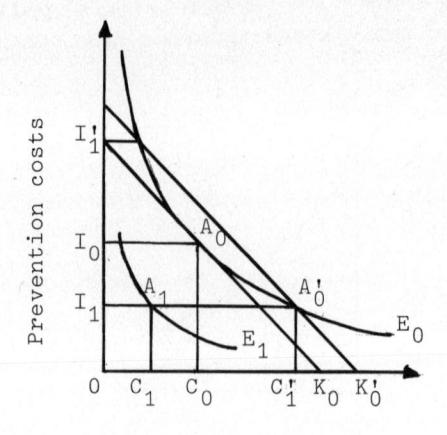

Figure 2

at A_1, paying prevention costs I_1 and restoration costs C_1. Viewed from a purely micro-economic standpoint, such behaviour is optimal for these firms, but it is inadequate from the social point of view.

Since the required level of environmental capacity is E_0 rather than E_1 at present scientific levels, the prevention remedies realized by the firms need to be combined with restoration costs C_1' instead of C_1 or C_0 and thus a part of these restoration costs will be transferred to the sufferers in the form of subsidies.

The unpaid part of restoration costs will be paid by the sufferers themselves. If they can not pay, physical losses will be incurred. Whether measurable or not, it is a fact that social costs in the shape of sufferers' expenditures or their physical losses occur in such a case. The greater the social costs are, the smaller will be the private costs to the firms. Hence in such a society profit-making firms and the general public are necessarily antagonistic towards each other. It is highly probable that Japan's success in her rapid growth was at least partly due to the sacrifice of necessary investments, either private or public, for the prevention of environmental disruption. We have found recently that in Japan strains on environmental capacity seem to be greater than those of heavy accumulation on production.

In the above-mentioned case the problem of unfair distribution of burdens due to environmental disruption has often been described as being a violation of "Polluters Pay Principles" but the problem of misallocation of resources due to lack of prevention is no less important. From the point of view of satisfactory levels of environmental capacity E_0, i.e. the ideal policy target, total costs for the society as a whole are higher than optimum costs : $K_0' = (I_1 + C_1') > K_0 = (I_0 + C_0)$. Thus the society as a whole makes unnecessary expenditure due to profit-maximizing objectives.

The isoquant E_0 will shift towards the origin with improvement in the technology of prevention and restoration, since the same level of environmental capacity will be realized at cheaper costs. Increases of industrial output will shift E_0 far from the origin, because maintenance of the same environmental capacity will cost more under the increased production.

Under conditions of given technology and industrial outputs the various kinds of environmental capacity curves - target curves - are usually situated in the following order starting from the origin and moving upwards and to the right : target curves of profit-making firms, governmental arbitration curves, curves representing inhabitant's or public demands and the ideal curves. In such a way, environmental capacity is not an exogenous variable but it also has a socio-economic character.

Some readers will doubt the adaptability of the isoquant curve to the reality of today's environmental problems. They may have the impression that prevention is always more effective than restoration and so the isoquant curve should not be applied to environmental problems. But this impression is nothing more than a reflection of reality. Today, in almost all countries of the world, methods of prevention of environmental disruption fall short of being adequate, so that a marginal increase in prevention can bring forth a large improvement in the environment and save tremendous amounts of restoration costs.

In the case of certain special kinds of production there seems to be no such environmental capacity. This would result in an isoquant at an infinite distance from the origin. Under these circumstances there is no choice but to prohibit such production.

Wolfgang Michalski (1965) who carried out an exhaustive study on various concepts of "social costs" once criticised William Kapp (1956), one of the pioneers in this field, saying that there was a lack of unity in his concepts of "social costs". According to Michalski there are four kinds of concepts of "social costs" in existing theories.

(1) Total costs which a product or project incur on a national economy.

(2) Losses due to divergence of a socio-economic system from its optimal condition.

(3) Non-marketable burdens on third persons or the community as a whole which are not taken care of by the firms responsible for these burdens.

(4) Implementation costs of economic policies.

Among the several different concepts used by Kapp which Michalski has pointed out, a more conspicuous lack of unity is said to exist between the second concept which Kapp employed, under the heading of "Sozialkosten", in Handwörterbuch der Sozialwissenschaften (Kapp, 1956), and the third concept mainly used in his "Social Costs and Private Enterprise" (Kapp, 1950).

According to Michalski, in that "Handwörterbuch", Kapp used the second of the above-mentioned four concepts, while be adapted the third concept in the "Social Costs and Private Enterprise". From the logical viewpoint represented in figure 2, this lack of unity in his concepts does not seem to be so fatal and contradictory. The above concepts are to be considered as the various facets of a many sided theoretical category. The third item of costs corresponds to $C_1 C_1'$ in figure 2, whereas the second item corresponds to $K_0 K_0'$ in the same figure.

The fourth concept, implementation costs of public policy, seems to be irrelevant to our logic of social costs, because it includes too many diverse items, e.g. implementation costs of military policy, education, urban development etc. none of which can always be regarded as social costs. At the same time we cannot say that public policy can cover all the necessary social costs, since it always falls short of the necessary prevention or restoration. Nevertheless it is not impossible to find a position for this concept on our figure or in our logic of social costs. If we define this concept as the necessary expenditure for implementation of public policy for prevention and restoration, it will be represented by certain parts of $I_1 I_1'$ and $C_1 C_1'$ (in figure 2).

It seems easier to identify the first concept, i.e. total costs which a product or project incur on a national economy, with another concept of social cost. It can be identified indeed with the concept of direct costs plus social costs of a product. But in fact identifying it in this way creates important problems.

It is interesting and useful for our discussion to find the name of Karl Marx among the economists cited by Michalski as those who employ the first concept. Michalski suggests that Marx's actual costs can be identified with the first concept of social costs. It is Michalsi's great interest in economic literature which has enabled him to apply such interpretations to past theories.

But the author can not readily agree with the identification of Marx's actual costs with the third concept of social costs. Marx's actual costs are represented by his famous c + v + m, where c is unvariable capital, v variable capital and m surplus value, whereas capitalists' cost consists of c + v only. But even Marx's actual costs of goods do not cover the necessary costs of prevention and restoration of environmental disruption. Hence the first concept of social costs contains more than Marx's actual cost at the level of the individual firm.

But Marx's concept of c + v + m is also used as an aggregate value for society as a whole. In this sense the <u>realized</u> prevention and restoration costs appear to be included in Marx's c + v + m. But still there remain <u>unrealized</u> prevention and restoration costs, viz. physical losses which are not only born by the enterprise responsible for them but also by the sufferers and public authorities.

III. <u>Policy Implications</u>

In the last part of section II we arrived at the frontier between economics and ecology. In the socio-economic system, the economic cycle either repeats itself or expands, replacing fixed capital and reproducing human labour capacity. If environmental capacities were replaced or reproduced by the same process at the same time, no problem would occur. In reality the economic cycle does not cover completely the disruption which it causes. Hence some parts of social costs take the form of physical losses, and are immeasurable in monetary terms, though they might be estimated to some extent in terms of shadow prices.

Even if we introduce government into our logic, the fundamental situation will not change very much. If the government pays prevention or restoration costs, i.e. $I_1 I_0 + C_1 C_0$ (figure 2), the optimal combination of remedies will be realized, but a distribution problem will be caused. If this policy is financed by taxes on polluting firms the Polluters Pay Principles can be adhered to. When they are financed by income taxes, these principles will be broken and environmental capacities will be violated at the expense of the standard of living. The real situation will lie somewhere between these two extremes and the distribution of costs ultimately depends upon the taxation system.

Mark-up pricing and inflation might be employed in order to transfer the social costs to the public. Thus, even if social costs are paid for by polluting firms, it will be possible to externalize them again through higher prices.

Consequently, through the mechanism of economic policy, the bills for social costs will be transferred continually to and from firms and the public in a market economy. Nevertheless, the more the P.P.P. is adhered to, the nearer the combination of policies will approach the optimal solution from both the point of view of an economically just allocation of costs and also from the ecological standpoint.

References

Michalski, Wolfgang., 1965, Grundlegung eines Operationalen Konzepts der Social Costs, (J.C.B. Mohr, Tübingen) pp. 6-13.

Kapp, William K., 1956, "Sozialkosten", in : Handwörterbuch der Sozialwissenschaften, Bd IX, (Gustav Fisher, Stuttgart).

Kapp, William K., 1950, Social Costs of Private Enterprise, (Schocken Books, New York).

Kapp, William K., 1963, Social Costs of Business Enterprise, (ASIA Publishing House, London).

THE THEOREM OF PUBLIC OVER-EXPENDITURE : THE ENVIRONMENTAL
OVERABATEMENT PROBLEM APPLIED TO BELGIAN HIGHWAY TRAFFIC

W. Desaeyere
Economische Hogeschool Limburg
Diepenbeek - Belgium

I. Introduction

The traditional theorem concerning overall public expenditure was formulated rather vaguely by Galbraith (1968, chapter XVII), stating that : "In general, society will consume too much private goods and not enough collective goods". Let us call this statement "the theorem of public underexpenditure".

This imbalance between the private and public sector of an economy is, according to Galbraith mainly, due to the free rider's phenomenon and the ensuing difficulties of raising funds that are necessary to finance the public goods.

It will be proven in section II that this theorem is in general not true and that actually the opposite may be the case (Desaeyere, 1975). The social imbalance may tip over in the other direction due to the absence of social cost pricing (section II.5) and the imposition of null standards (section II.6).

This danger of public overexpenditure is illustrated by the overall level of highway traffic in Belgium (section III).

Notation

e_i = level of a public parameter like degree of water pollution, congestion on the roads, etc.

x_j = quantity of a private good

v_g = quantity of a collective production factor like surface of purification station, length of the roads, etc.

P_j = price for x_j

TC_j = total production cost of x_j

TAC_g = total abatement cost corresponding with the collective production factor v_g

TEC_i = total environmental cost of society due to the public parameter e_i

SZ = social surplus

MC = marginal production cost

MAC = marginal abatement cost

MEC(x) = marginal environmental cost

MAB(v) = marginal benefit of abatement

$v_s(x)$ = abatement function

Conventions

- The upper indices of the current indices i, j and g are indicated by the corresponding capital letters I, J and G.
- The optimal level of a variable is indicated by the letter 0 above this variable.
- Four different decision rules are used : the social optimum, the private optimum, the private optimum with abatement and the null optimum. They are indicated by the indices S, R, RA and N respectively.

II. The theorem om public over-expenditure

1. The theorem and assumptions

The level of collective expenditure in an economy will in general be too high as a consequence of the absence of social cost pricing and as a consequence of environmental null standards[1].

The following assumptions are made :

a. Welfare of society is supposed to depend upon the consumption of private goods $x_j (j=1,\ldots,J)$ and the level of public parameters $e_i (i=1,\ldots,I)$.
Government has no direct impact on e_i. The level of public welfare can only indirectly be influenced by collective goods $v_g (g=1,\ldots,G)$.

b. The variables e_j, x_j and v_g and linked by the "environmental structure" which expresses the fact that the level of the public parameters e_i depend on the one hand upon the consumption of private goods x_j and on the other hand upon the collective goods v_g, which can in fact be seen as collective production factors that produce e_i. In general the "environmental structure" can be formulated as follows :

$$f_k(e_1,\ldots,e_i,\ldots,e_T,x_1,\ldots,x_j,\ldots,x_J,v_1,\ldots,v_g,\ldots,v_G) = 0$$

$$k = 1,\ldots,K$$

c. We will limit ourselves to a simple analysis by working directly with demand curves and cost functions[2]. This means that the following relations are supposed to be given :

- The demand curves for private consumption : $P_j = P_j(x_j)$
- The cost functions for the private goods : $TC_j = TC_j(x_j)$
- The cost functions for the collective production goods : $TAC_g = TAV_g(v_g)$
- The environmental cost functions : $TEC_i = TEC_i(e_i)$

d. The welfare of society can thus be represented by the social surplus SZ, the difference between the benefits and the costs of society :

$$SZ = \sum_{j=1}^{J} \{\int P_j(x_j)dx - TC_j(x_j)\} - \sum_{g=1}^{G} TAC_g(v_g) - \sum_{i=1}^{I} TEC_i(e_i) \quad (1)$$

2. A simplification

In order to simplify the analysis as much as possible we will reduce the dimensions by supposing that all sets consist of only one element or $J = I = G = K = 1$.
This means that the economy contains a private good x, a public parameter e and a collective production factor v and that the environmental structure reduces to one equation, the environmental function, which expresses the fact that the level of the public parameter e depends upon the private consumption x and upon the collective production factor v.

The model can thus be written as follows :

$$\text{Max !} \quad SZ = \int P(x)dx - TC(x) - TAC(v) - TEC(e) \quad (2)$$
$$x,v$$

$$\text{s.t.} \quad e = e(x,v) \quad (3)$$

3. Specification of the equations

The public parameter e is supposed to be a public bad i.e. the consumers prefer a lower level of e. This can be seen from the fact that the willingness to pay for e is introduced negatively, i.e. as a cost TEC, the total environmental costs. The parameter e can thus be seen as the degree of environmental deterioration : the degree of waterpollution, the degree of criminality etc.

As a consequence, the environmental function can be characterised by the following assumptions :

$\frac{\partial e}{\partial x} > 0$ i.e. an increase in private consumption x increases the level of environmental deterioration e.

$\frac{\partial e}{\partial v} < 0$ i.e. an increase in public expenditure for the collective production factor v reduces the level of environmental deterioration.

Of course, our choice is purely conventional : the opposite approach is perfectly possible. Indeed we could define e not as a public bad, but as a public good. This would oblige us to introduce a positive public demand curve for e, instead of an environmental cost function. Nothing essentially would be changed however, since compensating changes must then be made in the environmental function : $\frac{\partial e}{\partial x}$ would become negative and $\frac{\partial e}{\partial v}$ positive.

Anyway it must be stressed that the theorem does not follow from the specification choosen here.

The other equations have the usual form which means that the following requirements are supposed to be satisfied :

$\frac{\partial P}{\partial x} < 0$ i.e. the demand curve slopes downward.

$\frac{\partial TC}{\partial x} > 0$ i.e. the marginal production cost of consumption goods is positive

$\frac{\partial TAC}{\partial v} > 0$ i.e. the marginal cost of collective production or the marginal abatement cost is positive

$\frac{\partial TEC}{\partial e} > 0$ for $e > \overset{\circ}{e}^N$

$\frac{\partial TEC}{\partial e} = 0$ for $e < \overset{\circ}{e}^N$

The marginal environmental cost is positive in the interval $e < \overset{\circ}{e}^N$ but is zero as soon as the pollution is reduced to the point $\overset{\circ}{e}^N$, the null-environmental standard.

4. The social optimum

The optimization process leads to two optimum conditions:

1) Price-optimum

$$P(\overset{\circ}{x}{}^S) = MC(\overset{\circ}{x}{}^S) + MEC(\overset{\circ}{x}{}^S) \tag{4}$$

which is the famous rule that price should be equal to marginal social cost.

2) Abatement - optimum condition

$$MAB(\overset{\circ}{v}{}^S) = MAC(\overset{\circ}{v}{}^S) \tag{5}$$

which states that the marginal benefit of abatement should be equal to the marginal abatement cost, where:

$$MAB(\overset{\circ}{v}{}^S) = - MEC\,[e(\overset{\circ}{v}{}^S)]\frac{\partial e}{\partial v}$$

Especially the abatement-optimum condition is of interest here since from this optimum condition we can find the abatement function which specifies the optimum level of abatement for every possible level of private consumption:

$$\overset{\circ}{v}{}^S = v_S(x) \tag{6}$$

where of course:

$$\frac{\partial v_S(x)}{\partial x} > 0 \quad \text{i.e. more private consumption means more abatement} \tag{7}$$

The decisions concerning public expenditure may be wrong for two reasons: the price-condition may be violated and the abatement-condition may be neglected.

5. Case A : absence of social cost pricing

Absence of social cost pricing means that no environmental tax equal to the marginal environmental cost is levied.

We will suppose that the other optimum conditions are satisfied which means that price is put equal to marginal cost:

$$P(\overset{\circ}{x}{}^{RA}) = MC(\overset{\circ}{x}{}^{RA}) \tag{8}$$

We will call the solution obtained the private optimum with abatement.

$$\overset{\circ}{p}{}^{RA} < \overset{\circ}{p}{}^{S} \tag{9}$$

If we compare the social optimum condition (4) with the private optimum condition (8) and we notice that MEC(x) is positive for all x in a non-trivial case according to the assumptions stated, we may conclude that (9) holds, i.e. the private price will be lower than the social price.

Furthermore

$$\overset{\circ}{x}{}^{RA} > \overset{\circ}{x}{}^{S} \tag{10}$$

This inequality follows directly from (9) since the demand curve is assumed to slope downward.

Finally,

$$\overset{\circ}{v}{}^{RA} > \overset{\circ}{v}{}^{S} \tag{11}$$

This inequality follows directly from (10) since the abatement function is a positive relationship between x and v according to (7). Hence, it may be concluded that the absence of social cost pricing leads to overconsumption and thus to overabatement.

6. Environmental null standards

Sometimes, government not only neglects the price-optimum condition (4) but also the abatement-optimum condition (5) and imposes environmental standards, i.e. government tries to eliminate pollution completely. This means that $\overset{\circ}{e}$ is choosen in such a way that there are no more remaining environmental costs.

We will call the solution obtained, the null-optimum. It is defined as follows :

$$MEC(\overset{\circ}{e}{}^{N}) = 0 \tag{12}$$

and

$$\overset{\circ}{e}{}^{N} < \overset{\circ}{e}{}^{RA} < \overset{\circ}{e}{}^{S} \tag{13}$$

These inequalities follow directly from the fact that in a non-trivial case MEC(e) > 0 in the interval $e > \overset{\circ}{e}{}^N$.

It also follows that :

$$\overset{\circ}{v}{}^N > \overset{\circ}{v}{}^{RA} > \overset{\circ}{v}{}^S \qquad (14)$$

The level of abatement can be found by substituting $\overset{\circ}{e}{}^N$ and the level of x, which may be $\overset{\circ}{x}{}^S$ or $\overset{\circ}{x}{}^{RA}$, in the environmental function :

$$\overset{\circ}{e}{}^N = e(\overset{\circ}{x}{}^{RA} \text{ or } \overset{\circ}{x}{}^S, \overset{\circ}{v}{}^N)$$

from which $\overset{\circ}{v}{}^N$ can be derived :

Since $\frac{\partial e}{\partial v} < 0$ the inequalities (14) follow from the inequalities (13).

It may also be concluded that the imposition of null-environmental standards leads to overabatement.

III. An example : highways in Belgium

1. A model for Belgium in 1975

Symbols in this example are :

x = number of highway kilometer-cars (private good)
v = number of kilometers highway (public production factor)
e = highway traffic density expressed as number of cars/kilometer (public parameter)
t = time needed in order to drive one kilometer (expressed in minutes)
w = value of one minute

The following equations are assumed to exist :

Demand for highway traffic :

$$P = 4.5 - 0.000\ 000\ 05\ x \qquad (15)$$

Environmental function :

$$e = \frac{x}{v} \qquad (16)$$

Environmental cost function :

[2.3.] $t = 0.5$ $\qquad e \leq 10.000$ (17)

$\qquad t = 0.5 + 0.000\ 01\ (e - 10.000) \qquad e > 10.000$

[2.4.] $TEC = w\ t\ x$ (18)

[2.5.] $TAC = 20.000\ v$ (19)

Equation (15) implies that at the current price of 3 F highway traffic x will be 30 million kilometer-cars. Let us estimate the car-stock at 3 million and the number of kilometers driven a day per car at 66.66 kilometer[3]).

The total number of kilometers driven per day can ten be estimated as 200 million kilometers. If we suppose that 15 percent of the traffic uses highways, we obtain indeed the estimate of 30 million highway kilometer-cars a day. A reduction of the price to 2.5 F would increase traffic demand up to 40 million.

Equation (16) is a simple identity which says that traffic density is obtained by dividing traffic through the number of highway kilometers. If for example, there is a consumption of 30 million highway kilometer-cars a day and there are 1.500 kilometers highway, there is a traffic density of 20.000 cars a day, which is again a good approximation of the present situation.

Equation (17) represents a typical pollution phenomenon : the traffi density increases the time necessary in order to travel one kilometer. Notice that the null-standard equals 10.000 : once the density is reduced so far, average velocity is supposed to attain its maximum of 120 km/hour which means that the time needed to travel one km - the reciprocal of velocity - is equal to 0.5 minutes/km. When density increases to 20.000 cars/day, the avarage velocity is supposed to decline to 100km/hou which implies that 0.6 minutes are needed in order to travel one highway km.

Equation (18) is again a simple identity which says that the total environmental costs of society are equal to the time (t) needed to travel one km multiplied by the value of none minute (w) and by the consumption level i.e. the number of kilometers (x).

Notice also that the good considered here, viz. "traffic", is a club-good, which means that only those consuming the good are supposed to suffer from the pollution. The size of the affected population is not fixed but depends itself on the level of the consumption[4].

In order to simplify matters we will suppose that all the traffic costs are proportional to time. The value of time we will therefore be fixed rather high at 300 F per hour.

Equation (19) implies that the construction costs of one km highway is equal to 83.701.800 F i.e. it approaches 100 million francs, which is a reasonable estimate. If we accept as social discount rate of 6 percent and an average lifetime of a highway of 20 years, the annual equivalent capital cost amounts to 7.3 million francs a year <u>or 20.000 F a day</u>.

The model can be reformulated by substituting in (18) t by its value from (17) and e by its value from (16) :

$$TEC = 5 [0.5 + 0.000\ 01\ (\frac{x}{v} - 10.000)]\ x$$

2. Scenario 1 : The Galbraithian situation of social imbalance

Galbraith supposes either that government does not realise that there is any pollution or that government is not willing to interfere because priority is given to private consumption. This situation can be characterised by two assumptions.

The first assumption states that government is not willing to allocate ressources in order to abate the pollution phenomenon. The number of kilometers of highway remains zero.

The second assumption states that government does not levy environmental taxes in order to restrict private consumption. According to Galbraith an overwhelming importance will be attached to private goods.

We can formulate Galbraith's idea mathematically by supposing that each consumer only bears his own cost, viz. the average environmental cost, which replaces the marginal production cost in the case of club goods (Buchanan, 1965).

The first hypothesis leads to :

$$\overset{\circ}{v}{}^R = 0 \qquad (22)$$

The second hypothesis implies that x is fixed by the private optimum condition :

$$P(\overset{\circ}{x}{}^R) = AEC(\overset{\circ}{x}{}^B) \qquad (23)$$

where $AEC(x) = \dfrac{TEC(x)}{x} = 2 + 0.000\ 05\ \dfrac{x}{v}$

We obtain in our numerical example :

$$4.5 - 0.000\ 000\ 05\ x = 2 + 0.000\ 05\ \dfrac{x}{v}$$

Substituting the value v = 0 this reduces to :

$$4.5 - 0.000\ 000\ 05\ x = 2 + \infty$$

$$\overset{\circ}{x}{}^R = 0 \qquad (24)$$

So, when almost no highways are built, the traffic density is infinitely high and so are the costs : as a result highway traffic drops of course to zero. Private consumption is <u>choked off</u> completely by the pollution phenomenon. In a certain sense this is an extreme example of Galbraithian social imbalance. Yet this was the real situation in Belgium in the fifties : no highways were built and the value of traffic was low because of the congestion phenomenon.

3. Scenario 2 : The social optimum

Before studying the present situation we will immediately give the ideal situation. This is obtained by maximising the social surplus :

$$\underset{x,v}{Max!}\ SZ = \int_0^x P(x^c)\ dx^c - TEC - TAC \qquad (25)$$

$$= \int_0^x (4.5 - 0.000\ 000\ 05\ x)\ dx - (2x + 0.000\ 05\ \dfrac{x^2}{v})$$

$$- 20.000\ v$$

The first order conditions imply :

$$P(\overset{\circ}{x}{}^S) = MEC(\overset{\circ}{x}{}^S) \qquad (26)$$

and

$$MEC(\overset{\circ}{v}{}^S) = MAC(\overset{\circ}{v}{}^S) \tag{27}$$

This gives us the following system of non-linear equations :

$$4.5 - 0.000\,000\,05\,x = 2 + 0.000\,1\,\frac{x}{v} \tag{28}$$

$$-(-0.000\,05\,\frac{x^2}{v^2}) = 20.000 \tag{29}$$

The solution of which is :

$$\overset{\circ}{x}{}^S = 10.000.000 \tag{30}$$

$$\overset{\circ}{v}{}^S = 500 \tag{31}$$

In the optimal situation, 500 km highways will be built, traffic will be 10 million kilometers/cars a day and traffic density will be 20.000 cars a day. This situation is of course better than the Galbraithian situation of social imbalance. Wether we will obtain this ideal in practice, is not sure as the following scenarios will make clear.

4. Scenario 3 : Absence of price policy, optimal abatement

In practice, government will in almost all cases limit its intervention to an abatement policy without adopting however a price policy. Equation (27) can thus be retained, but equation (26) should be replaced by the private optimum condition (23). In reality, the optimum conditions are more complex in the case of a club good since AEC contains e which depends upon v. An approximate solution to the problem will be presented here.

The following system of non-linear equations is obtained :

$$4.5 - 0.000\,000\,05\,x = 2 + 0.000\,05\,\frac{x}{v} \tag{32}$$

$$-(-0.000\,05\,\frac{x^2}{v^2}) = 20.000 \tag{33}$$

The solution of which gives :

$$\overset{\circ}{x}{}^{RA} = 30.000.000.000 \tag{34}$$

$$\overset{\circ}{v}{}^{RA} = 1.500 \tag{35}$$

Now, 1.500 km highways will be built, which is three times the optimal quantity. Traffic will be 30 million kilometers/cars a day which implies again a traffic density of 20.000 cars a day. This corresponds approximately to the situation in Belgium in 1975.

5. Scenario 4 : Absence of price policy, environmental null standard

As we have seen in section III.4. the introduction of an abatement policy without an environmental tax leads to an overshooting of the objective and too many highways will be built. In spite of this, government is planning to build still more highways in Belgium. This error is due to the fact that government tries to eliminate all pollution and thus imposes an environmental null standard.

The level of v is choosen as follows : (t = 0.5)

e = 10.000 which leads to :

$$TEC = 2.5 \ x \tag{36}$$

and thus

$$AEC = 2.5 \tag{37}$$

Since the price policy is supposed to be absent, the private optimum condition (23) is used or :

$$P(\overset{\circ}{x}{}^N) = AEC(\overset{\circ}{x}{}^N) \tag{38}$$

or

$$4.5 - 0.000\ 000\ 05\ x = 2.5$$

or

$$\overset{\circ}{x}{}^N = 40.000.000 \tag{39}$$

Substituting this value into (16) gives :

$$\overset{\circ}{e}{}^N = \frac{40.000.000}{v} \tag{40}$$

Since $\overset{\circ}{e}{}^N$ must be kept at the level 10.000 this means :

$$\overset{\circ}{v}{}^N = 4.000 \tag{41}$$

In the case of the environmental null standard the pollution phenomenon will be completely eliminated : traffic density will remain at its minimum level of 10.000. This will cause traffic to rise to 40 million kilometers-cars : it will therefore be necessary to build 4.000 additional highway kilometers.

6. Conclusions

Scenario 3 is very attractive for the government : private consumption is allowed to treble whereas collective consumption also trebles. Yet this policy in fact reduces welfare from a benefit per day of 2.5 million F to a loss of 11.25 million F per day.

One may wonder how society will choose a solution where welfare goes down. This paradox is explained by the private myopia : each driver bears in fact a cost that is lower than the benefits be reveices. Indeed for all $x < \overset{o}{x}{}^A$, the willingness to pay $P(x)$ exceeds the price that will be paid : $AEC(\overset{o}{x}{}^A) = 3$ F per km. The abatement expenditures, i.e. the construction outlays of the highway, are not taken into account.

It may also be noted that abatement does not eliminate the pollution phenomeon. As well in scenario 2 as in scenario 3, a certain level of pollution persists : this co-called "residual pollution" chokes off some consumption and helps in reducing private consumption and thus pollution.

This finding also explains scenario 4. In this scenario there is no more residual pollution. Congestion is completely eliminated, and this result is very appealing to many politicians, because then the impact to the abatement policy becomes clearly visible for everyone. Yet this policy is wrong : social surplus declines still more. The loss now amounts to 40 million per day. In spite of this a further expansion of the highway system in Belgium is proposed by government.

Instead of expanding the highway system, a certain reduction of traffic should be our goal. This proposed reduction of traffic can be e.g. obtained by introducing an environmental tax on gasoline. In the example given, the price per km should rise from 3 F to 4 F per km.

Number of scenario	1	2	3	4
Name of scenario	Galbraithian situation of social imbalance (Private optimum)	Social optimum	Absence of price policy optimal abatement (Private optimum with abatement, approximate solution)	Absence of price policy, environmental null standard (Null optimum)
Subindex characterising the scenario	R	S	RA	N
Traffic, x (in km-cars highway)	0	10.000.000	30.000.000	40.000.000
Length of highways, v (in km)	0	500	1.500	4.000
Traffic density, e (cars per day/km)	∞	20.000	20.000	10.000
Time for 1 km, t (in minutes)	∞	0.6	0.6	0.5
Cost per car per km AEC = $\frac{TEC}{x}$ (in francs)	∞	3	3	2.5
Marginal cost MEC (in francs)	∞	4	4	2.5
Price (in francs)	∞	4	3	2.5
Environmental costs (10^6)	8	30	90	100
Abatements costs (10^6F)	0	10	30	80
Social costs (10^6F)	8	40	120	180
Total benefits (10^6F)	0	42.5	108.75	140
Social surplus (10^6F)	−∞	2.5	−11.25	−40

Table 1 : Comparison of the scenarios for highway construction in Belgium

Footnotes

1. The model used here is based on the work of S. Kolm (1968). In a narrow sense, the model only applies if the population, affected by pollution, remains fixed. Furthermore, the theorem approximately holds for "club goods".

2. This limitation is standard practise since Williamson's contribution (Williamson, 1966).

3. This implies that an average car drives approximately 23.000 kilometers a year, which is above the real annual mileage in Belgium.

4. An alternative formulation can be found in Winch (1973).

References

Buchman, J.M., 1965, An Economic Theory of Clubs, Economica, 1, 1-14.

Desaeyere, W., 1974, Un modèle urbain simple, in Proceedings of the World congress of the International Federation for Housing and Planning, Goas for Urban Development, Yesterday, Today, Tomorrow, (I.F.H.P., The Hague), pp. 409-492.

Desaeyere, W., 1975, Het sociale optimum bij congestie, in Het huis staat in brand. Bevolking en bevuiling, Eds., P. Van Moeseke, K. Tavernier (Standaard Wetenschappelijke Uitgeverij, Antwerp).

Galbraith, J.K., 1958, The Affluent Society (Haighton Mifflin, Boston).

Kolm, S.Ch., 1968, La théorie économique générale de l'encombrement (S.E.D.E.I.S., Paris).

Williamson, O.E., 1966, Peak-load Pricing and Optimal Capacity, American Economic Review, pp. 810-827.

Winch, D.M., 1973, The pure theory of non-pure goods, Canadian Journal of Economics, pp. 143-163.

LIMITATIONS OF REGIONAL AND SECTORSPECIFIC ECONOMIC GROWTH BY POLLUTION RESTRICTIONS AND SCARCITY OF RAW MATERIALS
- A REGIONALIZED MULTISECTOR MODEL -

G. Rembold
University of Karlsruhe
Karlsruhe - W. Germany

I. Introduction

During the past few years, especially two facts have gained a central position both in the theoretical and political discussions on economic growth : firstly, the increasing pollution effects exerted by the emission of poisonous and waste materials and by energy release on the potentials of air and water, on the climate, the flora and fauna, and last but not least, on mankind; secondly, the ever-increasing exploitation of industrial raw materials despite the limitations - even on a worldwide scale - of such resources.

The environment is widely affected, considerably by industrial production and partly by private consumption, to a degree which, in case of certain substances, has already outranged the limits of the regenerating capacities of nature itself. The remarkable regional differences in the emission of substances affecting the environment correspond to the high local differences in the concentration of industrial and consumptive activities. Thus, pollutant-concentration rates for the atmosphere within agglomeration centers exceed the rates measured in rural areas by far.

In addition, these locally very different environmental strains are accompanied by varying potentials of pollutant decomposition and, hence absorption as a result of geographical and meteorological conditions. Thus, regions frequently subjected to inversive weather conditions will be much more vulnerable than regions with a high air-circulation average; with regard to natural waters and rivers, the degree of pollution normally increases along the direction of flow. Consequently, statements concerning pollution limits not to be exceeded must take into account such regional differences.

Another problem faced in establishing maximum permissible pollution rates arises from the hardly clarified interdependence between emission as the cause of pollution and immission as the actual pollution influence. Let us base the following on the assumption that this interdependence has been clarified in that definitive maximum permissible emission rates can be determined for the individual pollutants and for energy release.

If such region-specific maximum rates of emmission are available and have a legal bearing, then there will be two consequences of importance to the economic development : firstly, increased efforts are required to avoid the emission of pollutants, in particular by installing the appropriate protective facilities such as filtration or purification plants, and secondly, restrictions have to be levied on those production processes where the emission of pollutants exceeds certain bounds despite the existence of protective facilities.

As a consequence, the effects that restrictions of pollutant emission and waste- and energy release have on the national economic development, must be analyzed with the aid of a regionally and sectorally disaggregated model; this will adequately illustrate the regional differences in the capacity of pollutant absorption on the one hand and the sectoral variations in pollutant emission on the other hand.

Furthermore, the model formulated as an attempt to take into account the consequences resulting from the augmenting scarcity of certain raw materials, as relevant in the sectoral economic development. Discussion about the restrictions of economic growth as determined by limited raw material resources has reached broader public interest, in particular due to the striking prognoses published by the Club of Rome (Meadows, 1972); and it first "concretization" was effected by the 1973 raw material crisis which brought about a variety of worldwide drastic price increases for important industrial raw materials. Though this crisis was primarily of a political nature, it also shed a warning light on the dependence of technologically advanced economies upon an adequate and secured system of raw-material supplies.

Within Europe, this dependence is underlined by the fact that most of the European countries have to compensate by imports their own shortage of raw materials. In case of extreme global price increases for raw materials, no matter whether they are due to factual or politically manipulated scarcity, a redistribution of incomes in favor of countries with abundant resources will result in the short run. In the long run, however, this redistribution will be accompanied by balance of payments problems, which must inevitably be followed by actions of economizing on imported raw materials by the countries with less abundant resources.

The regionalized multi-sector model described here is an effort to combine consideration of the "traditional" determinants of growth : factor equipment and factor productivity, technologic progress and production interrelations, investment- and consumption-demand behavior as well as infrastructure, with those consequences to the regional sector-specific economic growth that are effected by pollution restrictions and the scarcity of imported raw materials. Herewith, the model is supposed to render possible, not only a strictly theoretical analysis of the structure of the economic process, but also an application to the empirical description and explication of concrete economic developments. According to these two most important claims, the model should fulfill the following conditions :

1. Model relationships should be formulated in such a way as to make them numerically determinable by means of known econometric methods.
2. The scope of statistical-empirical information required for the numerical determination of the model parameters should be kept within realistic bounds.
3. Evaluation of the model should be possible with currently practicable mathematical methods that can be covered by computer techniques.

Only few of the models developed in growth theory so far take into account the spatial dimension, and only few of them meet the claims of an economic application. This is in particular due to the fact that in none of these models all of the essential real factors determining economic growth are integrated and thus quantifiable. Rather, most of the growth models - no matter whether of the Keynesian, neoclassic, or von-Neumann type - are representative of pure theory. This shows how urgent it is to develop dynamic models relevant to planning and enabling us to quantify the paths of growth of individual regions and sectors, at least approximately.

The economic interdependence between regions has to be considered as the substantial element of a spatially disaggregated model. It appears feasible to apply such model versions that already take into account aspects of interrelationships. The Input-Output Theory developed by Leontief (1951) proves to be the type of model most likely to correspond with the stated requirements. This theory offers an analysis that is first of all in accordance with the requirements of economic policy, its results being checkable by statistical methods.

The input-output concept, however, turns out to have a serious insufficiency in its dynamic version, that is, the model structure considers the equilibrium of the goods market as given "by assumption" rather than by fulfillment of a behavior-related condition of equilibrium. This is especially aggravating in market-economies where there is no guarantee that supply and demand patterns in all sectoral goods markets are in agreement to the extent that in these markets the factual supply and demand quantities are identical for each time period under consideration. Due to this lack, the dynamic input-output model along with the assumption of long-period full employment of capacities is, for arbitrary parameter configurations, not consistent with the basic claim for non-negative sectoral gross production rates (Solow, 1959; Rembold, 1975). This inconsistency can be avoided by following a suggestion put forth by Dorfman, Samuelson, and Solow (1958), in that the dynamic input-output model is changed into a linear dynamic optimization model, removing the assumptions of equilibrium and of long-period full employment of capacities.

For the model described here, this modified version of a dynamic input-output model has been chosen as a formal basic structure. However, due to the claims that numerical evaluation must be performable by computer techniques, the need arises, to change the formal mathematical structure in such a way as to formulate the model as a simultaneously solvable linear optimization model with a limited optimization horizon.

II. The Model

The model is formulated for an economy subdivided into m regions. The economy of each region consists of n production sectors to be constructed according to the procedure applied in Input-Output theory.[1] The primary inputs comprise, aside from capital (K_{ij}^s) and labor (A_j^s), the productively effective infrastructure (St^s) and h "scarce" types of industrial raw materials ($R_1, \ldots, R_h$).[2] This deviation from procedures applied in the Input-Output theory is justified especially in countries lacking raw materials, because here these raw-material supplies are widely covered by imports and, hence, the delivery of such raw materials is not an integral component of the domestic production structure.

Also included are private households, local, regional, and national public institutions and foreign countries, with all connected economic activities.

The k types of pollution and energy emission connected with production and consumption ($P_1, \ldots, P_k$) are interpreted as being "undesirable" outputs of those sectors where such substances originate.

The actual sector-specific production volume in the form of the factually realized gross value of production is the indicator variable for the economic development in the individual regions.

The most important symbols used in the model are:

1. Endogenous Variables:

A_i^s : total supply of type i goods (i = 1, ..., n) in region s (s = 1, ..., m)

B_i^{rs} : delivery flow from region r (r = 1, ..., m) to region s, of type i goods

C_i^s : private consumption of type i goods in region s

D_i^s : total demand for type i goods, in region s

1_{ij}^s : net investment demand of sector j in region s, for type i goods

$1P_{ij}^s$: demand for emission protection facilities in sector j of region s, produced in sector i

K_{ij}^s : capital stock of type i investment goods, in sector j of region s

P_{fi}^r : pollutant emission of type f (f = 1, ..., k) originating from sector i in region r

R_{gj}^s : consumption of "scarce" raw material g (g = 1, ..., h) in sector j of region s

St_j^s : input of productively effective regional infrastructure in sector j of region s

$_{pr,v}Y^s$: income available to private households in region s

X_{ij}^{rs} : current inputs (including reinvestments) in sector j of region s delivered by sector i in region r

X_j^r or X_j^s : gross value of production of sector i (j) in region r (s)

2. Exogenous Variables

G_i^s : demand of local and/or regional and/or national public institutions in region s for type i goods

HBS_i^s : balance of exports over imports of region s regarding type i goods ($EX_i^s - IM_i^s$)

$\text{aut}I_{ij}^s$: autonomous component of investment demand of sector j in region s for type i goods

St^s : infrastructure in region s

W_{ij}^{rs} : interregional and/or intersectoral migration of labor

W_j^s : labor input of sector j in region s

Z_i^s : demand for consumption goods of type i in region s, as independent from income

3. Parameters and Coefficients :

a_{ij}^s : input coefficient

b_{ij}^s : capital coefficient

c_i^s : marginal rate of consumption

d_i^{rs} : trade coefficient

es_j^s : quota of anti-emission costs per unit of output

p_{fi}^r : emission coefficient

q_{fi} : investment coefficient for anti-emission facilities

r_{gj}^s : coefficient of raw material inputs

j_{ind}^t or t_{dir}^s : average rates of indirect or direct taxes, respectively

$(1+u_i^s)^t$: growth rate of governmental demand for type i goods in region s

$(1\pm\bar{v}_j^s)^t$: rate of trade balance changes

w_j^s : labor coefficient

$(1\pm\alpha r_g)^t$: rate of type g raw-material scarcity changes

β_{ij}^s : marginal rate of investment

$(1\pm\gamma^s)^t$: rate of change in Z_i

δ_j^s : sector-specific coefficient of infrastructural inputs

$(1\pm\mu_j^s)^t$: rate of change in the intra-sector, intra-region number of manhours

$(1\pm\pi_{fC}^r)$: rate of change in pollutant emission within the consumption sector

ηs : quota of public expenditures improving the infrastructure

τ : symbol for technical progress

4. General Symbols :

$\hat{\ }$: symbol for vectors

$\hat{E}$: unit matrix

$\hat{P}$: permutation matrix

$\forall$: "for all"-operator

The total model according to Figure 1 is divided into four components : a production model, an environmental model, a demand model, and a trade model.

1. Production Model

Goods and services ($X_j^s(t)$) produced during each time period t (t = 1, ..., T), within each sector j of a region s, are determined by : existing capital stock ($K_{ij}^s(t)$) and number of manhours ($W_j^s(t)$) as well as their utilization rate by the realized technical progress ($V^\tau{}_{ij}^s$, $K^\tau{}_{ij}^s$, $R^\tau{}_j^s$, $W^\tau{}_j^s$); by the amount of current inputs ($X_{ij}^{rs}(t)$); by the regional infrastructure equipment ($St^s(t)$); by the supply of raw materials in the national economy ($\bar{R}_g(t)$); and by the limitations levied on the emission of pollutants.

Based on the fundamental assumption of input-output theory with regard to production, it is assumed that the current as well as the primary inputs of capital, labor, and productively effective infrastructure can be approximated, at least for several periods, by linear homogenous, limitational input functions. Production-determining effects of technical progress are taken account of in a first approximation, by the time variant input coefficients. This is based on the assumption

PRODUCTION MODEL

- labor potential
- technical progress
- limitation of raw materials
- capital stock
- **regionalized sector-specific production**
- current inputs
- regional infrastructure equipment
- anti-emission facilities/maximum permissible emission rates

TRADE MODEL

region-specific supply of goods and services

region-specific demand of goods and services

ENVIRONMENTAL MODEL

emission of pollutants and of energy

region-specific maximum permissible emission rates

DEMAND MODEL

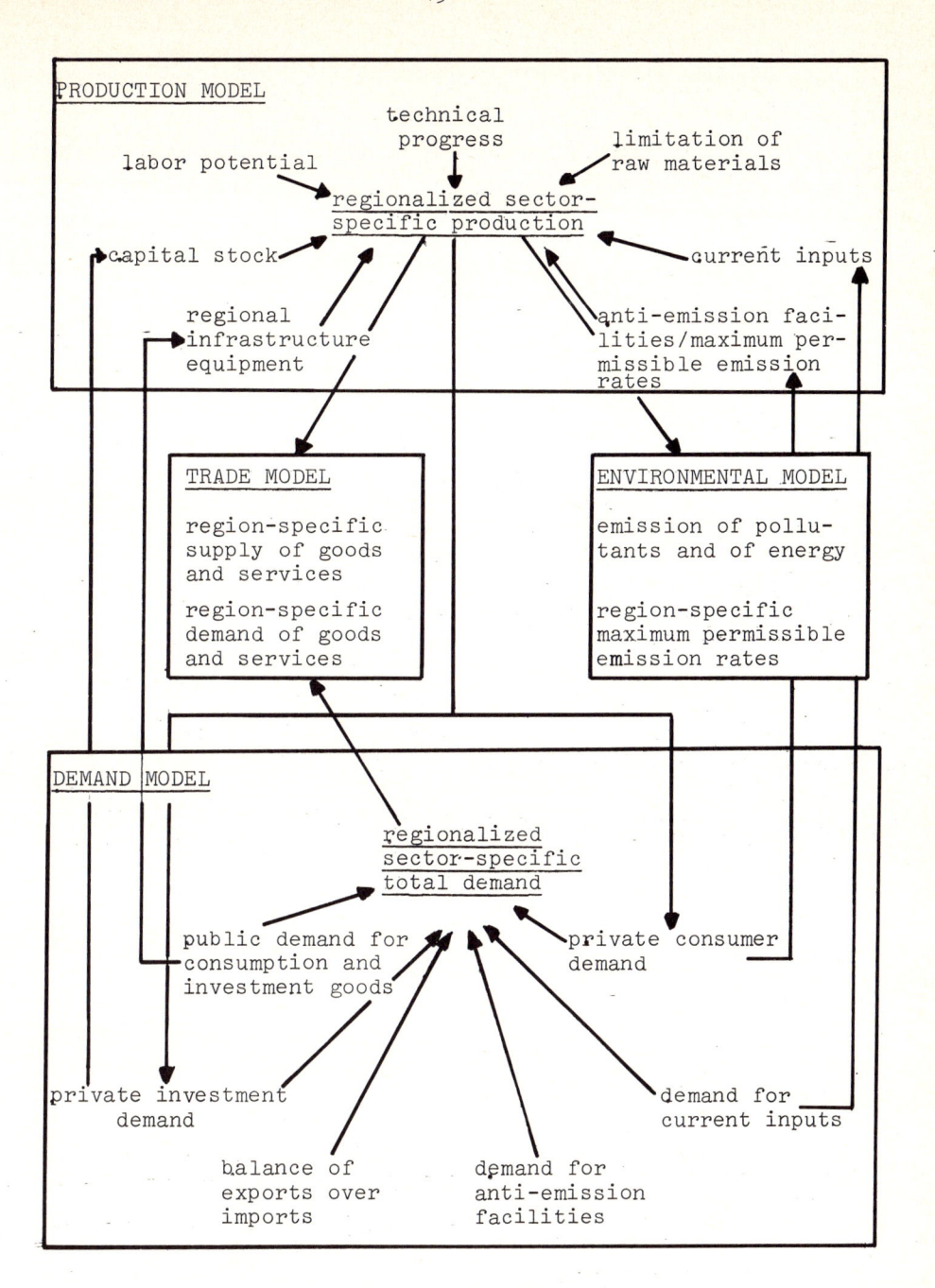

regionalized sector-specific total demand

- public demand for consumption and investment goods
- private consumer demand
- private investment demand
- balance of exports over imports
- demand for anti-emission facilities
- demand for current inputs

Figure 1 : Structure of the Model

that the effects of technical progress consist of, firstly an increase in the productivity of labor and capital and, secondly, the changes in the amount of current inputs within a certain framework. For the input functions, this yields the following:

$$X_{ij}^{s}(t) = a_{ij}^{s} \cdot (1 \pm {}_{V}\tau_{ij}^{s})^t \cdot X_{j}^{s}(t) \tag{1}$$

$$K_{ij}^{s}(t) = b_{ij}^{s} \cdot (1 - {}_{K}\tau_{ij}^{s})^t \cdot X_{j}^{s}(t) \quad, \forall\ i,j,s,t, \text{ where} \tag{2}$$

$$W_{j}^{s}(t) = w_{j}^{s} \cdot (1 - {}_{W}\tau_{j}^{s})^t \cdot X_{j}^{s}(t) \quad 0 \leq {}_{V}\tau_{ij}^{s}, {}_{K}\tau_{ij}^{s}, {}_{W}\tau_{j}^{s} \leq 1 \tag{3}$$

$$St_{j}^{s} = {}_{j}^{s} \cdot X_{j}^{s}(t) \tag{4}$$

Let the consumption of "scarce" raw materials ($R_{gj}^{s}(t)$) in each sector j also be determined by linear-homogenous, limitational consumption functions:

$$R_{gi}^{s}(t) = r_{gi}^{s}(0) \cdot (1 - {}_{R}\tau_{gi}^{s})^t \cdot X_{j}^{s}(t) \quad, \forall\ j,g,s,t, \text{where} \tag{5}$$

$$0 \leq {}_{R}\tau_{j}^{s} \leq 1.$$

Herein, the time-variant raw-material input coefficients $(r_{gi}^{s}(0) \cdot (1 - {}_{R}\tau_{gi}^{s})^t$ are to indicate that through technical innovation, the dependence of sectoral production upon certain types of raw material can vary, for instance if a scarce type of raw material is used with a higher intensity than before.

Aside from statements concerning the inputs determined by production techniques, a dyanmic model requires corresponding statements concerning the intertemporal development of factor equipments.

The intertemporal variation of capital stock ($K_{ij}^{s}(t)$) is determined by the net investment demand ($I_{ij}^{s}(t)$) realized within each time period:

$$K_{ij}^{s}(t) = K_{ij}^{s}(0) + \sum_{\varepsilon = 1}^{t} I_{ij}^{s}(\varepsilon) \quad, \forall\ i, j, s, t. \tag{6}$$

It is assumed that the sector-specific labor supply in each region ($W_{j}^{s}(t)$) changes intra-regionally and intra-sectorally at a given rate ($(1 \pm \mu_{j}^{s})^t$) on one hand, and due to interregional and/or intersectoral migration ($W_{ij}^{rs}(t)$) on the other hand. Herein, migration values must be predicted exogenously within a migration model :

$$W_j^s(t) = (1 \pm s_j)^t \cdot W_j^s(0) + \sum_{\varepsilon=1}^{t} \sum_{r=1}^{m} \sum_{i=1}^{n} W_{ij}^{rs}(\varepsilon)$$

$$- \sum_{\varepsilon=1}^{t} \sum_{r=1}^{m} \sum_{i=1}^{n} W_{ji}^{sr}(\varepsilon) \quad , \forall j, s, t. \quad (7)$$

$$r \neq s \quad i \neq j$$

The region-specific infrastructure expands from one period to another, depending upon the expenditure behavior of public institutions (G_j^s) in region s :

$$St^s(t) = St^s(0) + \sum_{\varepsilon=1}^{t} \eta^s(\varepsilon) \cdot \sum_{j=1}^{n} (1+u_i^s)^t \cdot G_i^s(0), \quad \forall s,t. \quad (8)$$

Let the consumption of raw materials of type g on the national level be confronted with a limited raw-material potential ($_A R_h(t)$) for each period :

$$_A R_g(t) = (1 \pm \alpha_{r_g})^t \cdot {_A R_h(0)} \quad , \forall g, t, \text{ where } \quad (9)$$
$$0 < \alpha_{r_g} \leq 1.$$

Corresponding to variations in the "scarcity degree" of type g raw material, the national economic potential will diminish or increase from one economic period to another at a rate of $(1 \pm \alpha_{r_g})^t)$.

Goods and services produced within each period are depicted by the existing capital stock, labor supply, and productive infrastructure as defined in relationships (6), (7) and (8), and by the potential in "scarce" raw materials g as defined in relationship (9). Combined with the input functions (2), (3), (4), (5), equations (6), (7), (8) and (9) result in four systems of inequalities which, along with the effects of limited pollutant and energy emission to be formulated in the following section, determine the sectoral production capacities per period and per region :

$$K_{ij}^s(t) \geq (1 - _K\tau_{ij}^s)^t \cdot b_{ij}^s(0) \cdot X_j^s(t) \quad (10)$$

$$W_j^s(t) \geq (1 - {_W}\tau_j^s)^t \cdot w_j^s(0) \cdot X_j^s(t) \qquad , \forall\ i,j,s,g,t. \qquad (11)$$

$$St^s(t) \geq \sum_{j=1}^{n} \delta_j^s \cdot X_j^s(t) \qquad (12)$$

$$A_g^R(t) \geq \sum_{j=1}^{n} \sum_{s=1}^{m} (1 - {_R}\tau_{gi}^s)^t \cdot r_{gi}^s(0) \cdot X_j^s(t) \qquad (13)$$

The formulation chosen for relations (10) and (11) permits to take explicit account of the varying sector-specific employment of capital stock and labor potential of each region as dependent upon the market situation. The inequality (12) is based on the assumption that the regional infrastructure potential given in each period is utilized by all sectors of a region jointly but at different intensities. In system (13) finally, the supply of type g raw material on the national level delimits the consumption of raw material in each region.

2. Environmental Model

During the production process, a variety of "undesirable" environment pollutants is generated aside from the "desired" output ($X_i^r(t)$), and/or energy in the form of waste heat or noise is emitted. Assuming in first approximation that these emissions ($P_{fi}^r(t)$) are generated in proportion with the factual production volume, the following emission functions can be formulated:

$$P_{fi}^r(t) = p_{fi}^r \cdot X_i^r(t) \qquad , \forall\ i, r, f, t. \qquad (14)$$

In addition, emissions of environmental pollutants occur in private and public consumption, in particular in the transport sector and as a result of the heating of buildings. Due to an inelastic demand for fuel and gasoline, which represents the main cause of this type of pollution, it appears reasonable to introduce to this model the simplifying assumption that pollutant emissions of the consumption sector change at an autonomously given rate (($1+\pi_{fC}^r)^t$):

$$P_{fC}^r(t) = (1 + \pi_{fC}^r)^t \cdot P_{fC}^r(0) \qquad , \forall\ r, f, t. \qquad (15)$$

Then, without anti-emission facilities, the following quantities of pollutants and energy would be emitted in region r per period :

$$P_f^r(t) = \sum_{i=1}^{n} P_{fi}^r(t) + P_{fC}^r(t) \quad , \forall\, r, f, t. \quad (16)$$

If legal bounds for the maximum permissible emission rates for each region r ($_{max}P_f^r$) exist, then the following three consequences must be considered in addition : firstly, the establishment of such region-specific legal bounds for maximum permissible rates may result in the necessity to cut down on pollutant-generating production in certain regions, or to limit the growth of such production. This consideration holds, if the protecting facilities in such production platns would not provide an adequate reduction of emissions. The following system of restrictions is, hence, applicable to the non-reduceable types of emission f^* :

$$_{max}P_{f^*}^r \geq \sum_{i=1}^{n} p_{f^*i}^r \cdot X_i^r(t) + (1 + \pi_{f^*C}^r)^t \cdot P_{f^*C}^r(0),$$
$$\forall\, r, t, f^*, \quad (17)$$

where f^* denotes those types of emission that cannot be prevented sufficiently. Secondly, anti-emission facilities must be installed within the production sectors involved. This results in increasing resource requirements necessary to construct such facilities. Assuming that the demand of sector j in region r for such facilities from sector i ($IP_{ij}^s(t)$) will be approximately proportional to the pollutant emission considered :

$$IP_{ij}^s(t) = \sum_{\substack{f=1 \\ f \neq f^*}}^{k} qf_i \cdot P_{fj}^s = \{\sum_{\substack{f=1 \\ f \neq f^*}}^{k} qf_i \cdot p_{fj}^s\} \cdot X_j^s(t)$$
$$, \forall\, i, j, s, t. \quad (18)$$

The porportionality coefficient (qf_i) herein indicates the actual net investment demand of anti-emission facilities from sector i for each type of pollutant emission f generated in sector j of region s, needed to avoid pollution rates exceeding the region-specific maximum permissible values.

Thirdly, the real production costs in each sector j of each region s will increase by the costs expended for anti-emission facilities ($ES_j^s(t)$); therefore, for each output unit is a certain portion of real anti-emission costs, and this portion is assumed to be constant:

$$\frac{ES_j^s(t)}{X_j^s(t)} = es_j^s \qquad , \forall\ j,\ s,\ t \qquad (19)$$

3. Demand Model

The demand model functionally depicts the demand behavior of the individuals in region s for goods and services of type i.

It is assumed that the demand of the production-sector for current inputs is determined purely technologically by the corresponding input functions (1):

$$X_j^s(t) = \sum_{j=1}^{n} a_{ij}^s(0) \cdot (1 - {}_V\tau_{ij}^s)^t \cdot X_j^s(t) \qquad , \forall\ i,s,t \qquad (20)$$

The private-household demand for consumption goods of type i ($C_i^s(t)$) depends upon available private incomes (${}_{pr,v}Y^s(t)$), taking account of a basic consumption quantity (Z_i^s), independent of income:

$$C_i^s(t) = c_i^s \cdot {}_{pr,v}Y^s(t) + (1 \pm \gamma^s)^t \cdot Z_i^s(0) \qquad , \forall\ i,\ s,t \qquad (21)$$

The rate of change ($(1 \pm \gamma^s)^t$) takes into consideration the population development retraceable to natural changes and migration. The available private income within region s results from the regional gross value of production, subtracting the portion of real costs for current inputs, the costs of raw materials, the costs for emission prevention, and taking account of the average rates of direct and indirect taxes:

$$\begin{aligned}
{}_{pr,v}Y^s = (1 - f_{dir}^s) \cdot \sum_{j=1}^{n} \{1 &- \sum_{i=1}^{n} a_{ij}^s(0) \cdot (1 \pm {}_V\tau_{ij}^s)^t \\
&- \sum_{g=1}^{k} r_{gi}^s(0) \cdot (1 - {}_R\tau_{gi}^s)^t \\
&- es_j^s - {}_j t_{ind}\} \cdot X_j^s(t) \\
&\qquad\qquad , \forall\ s,\ t. \qquad (22)
\end{aligned}$$

Entrepreneurs are planning their demand for net investments ($I_j^s(t)$)

both autonomously and dependent upon the temporal development of the gross value of production. The reinvestments are assumed to be part of the current inputs :

$$I_j^s(t) = \sum_{j=1}^n \{\beta_{ij}^s \cdot [X_j^s(t) - X_j^s(t-1)] + {}_{aut}I_{ij}^s\} \quad , \forall\ i,s,t. \quad (23)$$

In equation (23), the parameters (β_{ij}^s) indicate the behavior-determined, production-specific marginal net-investment rates of sector j enterprises in region s.

Let the demand behavior of public institutions $(G_i^s(t))$ in region s show increases in all expenditure components with production-specific, autonomously given rates :

$$G_i^s(t) = (1 + u_i^s)^t \cdot G_i^s(0) \quad , \forall\ i,\ s,\ t \quad (24)$$

Let foreign trade be reflected by the demand model, e.g. by the net values of regional sector-specific export and import demands. The region-specific net figures of the trade-balances $(HBS_i^s(t))$, of the production sectors are trend-extrapolated on the basis of past rates; consequently, the following expression becomes formally applicable :

$$EX_i^s(t) - IM_i^s(t) = HBS_i^s(t) = (1 \pm v_i^s)^t \cdot HBS_i^s(0)$$

$$, \forall\ i,\ s,\ t \quad (25)$$

Using demand relations (18), (20), (21) in connection with relations (22), (23), (24) and (25) yields the demand for type i goods and services prevailing in region s. Written in vector form, the demand functions for the markets of all type i goods in region s can be expressed by the following relationship :

$$\hat{D}^s(t) = \hat{A}^s(t) \cdot \hat{X}^s(t) + \hat{c}^s(t) \cdot X^s(t) + \widehat{(1+\gamma)}^{s\ t} \cdot \hat{Z}^s(0) + \hat{\beta}^s$$

$$\{\hat{X}^s(t) - \hat{X}^s(t-1)\} + {}_{aut}\hat{I}^s + \widehat{(1+u)}^{s\ t} \cdot \hat{G}^s(0) + \widehat{(1\pm v)}^{s\ t} \cdot$$

$$\widehat{HBS}^s(0) + \hat{QP}^s \cdot \hat{X}^s(t) \quad , \forall\ s,\ t \quad (26)$$

where

$\hat{A}^s(t)$ = matrix of $a_{ij}^s(0)\ (1 \pm {}_v\tau_{ij}^s)^t$

$\bar{c}_i^s(t)$ = matrix of modified marginal rates of consumption :

$$c_i^s \cdot (1-t_{dir}^s) \sum_{j=1}^{n} \{1 - \sum_{i=1}^{n} a_{ij}^s(0) \cdot (1-_V\tau_{ij}^s)^t - \sum_{g=1}^{k} r_{gj}^s(0) \cdot$$

$$(1-_R\tau_{gj}^s)^t - es_j^s - _jt_{ind}\}$$

$\hat{\beta}^s$ = matrix of β_{ij}^s

$\hat{QP}^s$ = matrix of $\sum_{\substack{f=1 \\ f \neq f^*}}^{k} qf_i \cdot p_{fj}^s$

4. Trade Model

The total demand in region s for each period ($\hat{D}^s(t)$), is facing a total supply of type i goods ($\hat{A}_i(t)$) constituted by intraregional production and the availability of mobile goods coming in from all remaining regions. Within a variety of determinants for the spatial distribution of products, e.g., saptial supply-behavior of producers, spatial demand behavior of buyers, structure and capacity of the transportation system, and interregional distances, it is usually the interregional distances only that can be quantified.

According to a model suggested by Chenery (1956) and Moses (1955), trade coefficients d_j^{rs} are deduced on the basis of the ex-post values of interregional commodity flows; these trade coefficients reveal which part of the total supply in region s originates from the sector i of a region r :

$$B_i^{rs}(t) = d_i^{rs} \cdot A_i^s(t) \qquad , \forall \ i, r, s, t \qquad (27)$$

The summing up over all receiving regions s in equation (27) allows to state a relationship between regional subdivided gross values of production of sector i and all regional supplies of goods of the same sector :

$$\sum_{s=1}^{m} B_i^{rs}(t) = X_i^r(t) = \sum_{s=1}^{m} d_i^{rs} \cdot A_i^s(t) \qquad , \forall \ i, r, t \quad (28)$$

Finally, summarizing all of the r delivering regions in a vector equation yields a relationship between the regionally subdivided total supply of type i goods and - again subdivided regionally - the gross values of production of the sector producing this type of goods :

$$\hat{X}_i(t) = \hat{T}_i \cdot \hat{A}_i(t) \qquad , \forall\ i,\ t \qquad (29)$$

where $\hat{T}_i$ = matrix of the d_j^{rs}

5. Integration of the Four Partial Models into a Total Model

The trade model represented in relation (29) forms the link between the total demand exerted in the markets of region s during each period (equation (26)) and the systems of production restrictions (10), (11), (12), (13) and environmental restrictions (17) that determine the sectoral production of the individual regions. As initially emphasized, there is no guarantee for the demand and supply behavior in each market to correspond to a degree where observed supply and demand balance in each period; it is, therefore, assumed that equilibrium disturbances might occur at the regional sector-specific markets in such a way that demand does not cover the supply potential :

$$A_i^s(t) \geqslant D_i^s(t) \qquad , \forall\ i,\ s,\ t \qquad (30)$$

Taking account of a permutation-matrix $\hat{P}$, which makes the vectorial orders of vector equations (26) and (29) to be in correspondence, yields a market relation derived from equations (26), (29), and the inequality (30) :

$$\hat{X}(t) \geqslant \hat{P}^{-1} \cdot \hat{T} \cdot \hat{P} \cdot \{\hat{A}(t) + \hat{\bar{c}}(t) + \hat{\beta} + \hat{QP}\} \cdot \hat{X}(t) - \hat{P}^{-1} \cdot \hat{T} \cdot \hat{P} \cdot \hat{\beta} \cdot \hat{X}(t-1)$$
$$+ \hat{P}^{-1} \cdot \hat{T} \cdot \hat{P} \cdot \{(\hat{E}+\hat{\gamma})^t \cdot \hat{Z}(0) + _{aut}\hat{I} + (\hat{E}+\hat{U})^t \cdot \hat{G}(0) + (\hat{E}\pm\hat{V})^t \cdot \hat{HBS}(0)\}$$

$$, \forall\ t. \qquad (31)$$

Moreover, considering first the non-negativity restrictions for the sectoral gross value of production of each region,

$$X_j^s(t) \geqslant 0 \qquad , \forall\ j,\ s,\ t, \qquad (32)$$

and secondly - in connection with the irreversibility assumption of the capital formation process -, the system of restrictions

$$X_j^s(t) - X_j^s(t-1) \geq 0 \qquad , \forall\; j, s, t, \qquad (33)$$

the system of inequalities (10), (11), (13), (31), (32) and (33) describes a multitude of growht paths available for the regional sector-specific gross values of production $X_j^s(t)$ within a span of T time periods.

Now, with the aid of an optimization algorithm, this multitude of solution possibilities allows to select those growth paths that will find maximum regional and sector-specific gross values of production in view of the behavior, production techniques, resource potentials, and environmental restrictions on which the model is based. For this purpose, the system of inequalities must be supplemented by a linear objective function that reveals the highest possible national gross value of production for the final time period during the process of optimization :

$$Z = \sum_{t=1}^{T} \sum_{r=s=1}^{m} \sum_{i=j=1}^{n} \alpha_i^r(t) \cdot X_i^r(t), \text{ where } \alpha_i^r = \begin{cases} 1 \text{ for } t = T \\ 0 \text{ otherwise.} \end{cases}$$

(34)

Thus, the total model represents - in a formal mathematical way - a dynamic optimization model which due to the lack of operational algorithms for problems of this magnitude, has to be reformulated to get to a linear optimization model that can be solved simultaneously. Disintegrating the values $X(t)$ in inequality system (31), and substituting for reasons of simplifying expressions,

$$\hat{T}^* = \hat{P}^{-1} \cdot \hat{T} \cdot \hat{P}$$

$$\hat{F}(t) = \{\hat{E} - \hat{T}^* \cdot [\hat{A}(t) + \hat{c}(t) + \hat{B} + \hat{QP}]\}$$

$$\widehat{AUT}(0) = \{(\hat{E}+\hat{\gamma})^t \cdot \hat{Z}(0) +{}_{aut}\hat{I} + (\hat{E}+\hat{U})^t \cdot \widehat{HBS}(0)\},$$

yields the formulation of the linear optimization program :

$$\max Z = \sum_{t=1}^{T} \sum_{r=s=1}^{m} \sum_{i=j=1}^{n} \alpha_i^r(t) \cdot X_i^r(t)$$

with the following constraints :

1. Market Constraints

$$\hat{F}(1) \cdot \hat{X}(1) \geq -\hat{T} \cdot \hat{\beta} \cdot \hat{X}(0) + \hat{T}^{\star} \cdot \widehat{AUT}(1)$$

$$\hat{T}^{\star} \cdot \hat{\beta} \cdot \hat{X}(1) + \hat{F}(2) \cdot \hat{X}(2) \geq \hat{T}^{\star} \cdot \widehat{AUT}(2)$$

$$\vdots$$

$$\hat{T}^{\star} \cdot \hat{\beta} \cdot \hat{X}(T-1) + \hat{F}(T) \cdot \hat{X}(T) \geq \hat{T}^{\star} \cdot \widehat{AUT}(T)$$

2. Capital Constraints

$$\{(1 - {}_K\tau_{ij}^s)^1 \cdot b_{ij}^s(0) - \beta_{ij}^s\} \cdot X_j^s(1) \leq K_{ij}^s(0) - \beta_{ij}^s(0) + 1 \cdot {}_{aut}I_{ij}^s$$

$$\vdots$$

$$\{(1 - {}_K\tau_{ij}^s)^T \cdot b_{ij}^s(0) - \beta_{ij}^s\} \cdot X_j^s(T) \leq K_{ij}^s(0) - \beta_{ij}^s \cdot X_j^s(0) + T \cdot {}_{aut}I_{ij}^s$$

3. Labor Potential Constraints

$$(1 - {}_W\tau_j^s)^1 \cdot w_j^s(0) \cdot X_j^s(1) \leq (1 {}^+_- \mu_j^s)^1 \cdot w_j^s(0) + \sum_{r=1}^{m} \sum_{i=1}^{n} W_{ij}^{rs}(1)$$

$$- \sum_{\substack{r=1 \\ r \neq s}}^{m} \sum_{\substack{i=1 \\ i \neq j}}^{n} W_{ji}^{sr}(1)$$

$$\vdots$$

$$(1 - {}_W\tau_j^s)^T \cdot w_j^s(0) \cdot X_j^s(T) \leq (1 {}^+_- \mu_j^s)^T \cdot w_j^s(0) + \sum_{\epsilon=1}^{T} \sum_{r=1}^{m} \sum_{i=1}^{n} W_{ij}^{rs}(\epsilon)$$

$$- \sum_{\epsilon=1}^{T} \sum_{\substack{r=1 \\ r \neq s}}^{m} \sum_{\substack{i=1 \\ i \neq j}}^{n} W_{ji}^{sr}(\epsilon)$$

4. Constraints of Raw Material Availability

$$\sum_{j=1}^{n}\sum_{s=1}^{m} (1-_R\tau_j^s)^1 \cdot r_{gj}^s(0) \cdot X_j^s(1) \leq (1_-^+ \alpha_{r_g})^1 \cdot {}_A R_g(0)$$

$$\vdots$$

$$\sum_{j=1}^{n}\sum_{s=1}^{m} (1-_R\tau_j^s)^T \cdot r_{gj}^s(0) \cdot X_j^s(T) \leq (1_-^+ \alpha_{r_g})^T \cdot {}_A R_h(0)$$

5. Infrastructural Constraints

$$\sum_{j=1}^{n} \delta_j^s \cdot X_j^s(1) \leq St^s(0) + n^s(1) \cdot \sum_{i=1}^{n} (1+u_i^s)^1 \cdot G_i^s(0)$$

$$\vdots$$

$$\sum_{j=1}^{n} \delta_j^s \cdot X_j^s(T) \leq St^s(0) + n^s(T) \cdot \sum_{i=1}^{n} (1+u_j^s)^T \cdot G_i^s(0)$$

6. Environmental Constraints

$$\sum_{i=1}^{n} P_f \star_i^r \cdot X_i^r(1) \leq \max P_f \star^r - (1_-^+ \pi_f \star_C^r)^T \cdot P_f \star_C^r(0)$$

$$\vdots$$

$$\sum_{i=1}^{n} P_f \star_i^r \cdot X_i^r(T) \leq \max P_f \star^r - (1_-^+ \pi_f \star_C^r)^T \cdot P_f \star_C^r(0)$$

where $i = j$ and $r = s$

7. Additional Formal Constraints

$$X_j^s(t) - X_j^s(t-1) \geq 0$$

$$X_j^s(t) \geq 0$$

Constraints 2. through 7. are applicable $\forall\ i,j,r,s,t,f^\star,g$.

The linear optimization program is composed of
{m . [T.(n^2+4n+1)-n+f^*]+g.T} constraints with (m.n.T) variables. Its
pattern is block-triangular and thus allows its computation for extensive structures, using modified Simplex procedures (see for instance Danzig (1955)). Before applying Simplex procedures, one must define slack variables. They can be interpreted in case of market constraints, as unsold sectoral production, per time period; in case of capital constraints, of restrictions of raw-material availability, and of infrastructure constraints as unused capacity or materials, and finally, in case of labor potential constraints as unemployment.

The assumptions regarding behavior and production, on which the model is based, describe the optimization horizon to be kept within relatively narrow limits. It is only for short durations that acceptable approximations to the developments of the real-world economy can be expected. And it is only continuous adaptation of assumptions regarding behavior and production techniques to the changes resulting as the time passes, that will allow longer-range development prognoses.

Moreover, the extent of computation on one hand and of some model assumptions on the other hand require a sort of regionalization and of sectoral split up to meet the following conditions : since the regional consumption demand of private households is solely determined by the incomes generated within the region, and since the demand as such becomes effective only at the market of the same region, regions must be conceived in a way to allow commuter movements beyond regional boundaries to be minimized. This requirement, as a rule, will be met in spatially large regions, the number of regions to be distinguished decreasing at the same time accordingly. The number of production sectors to be distinguhished is substantially dependent upon the extent of available information, and finally the optimization horizon - based on the restrictive assumptions underlying the model - should not exceed 10 years.

III. Concluding Remarks

It has been shown by means of the framework described here that a formal attempt to take account of all substantial determinants of regional and sector-specific production and demand out of the many growth factors is feasible. It has been evidenced, e.g., that consumption limitation for certain kinds of raw material can be interpreted as additional primary inputs that limit the sectoral production potential, and that the limitation of emissions polluting the environment has production-restricting effects. However, unlike the environmental models developed in the past (e.g. by Thoss, 1973), it has been shown that legal maximum rates of emission do induce a corresponding demand for anti-emission facilities and, thereby induce an increase of production costs within the sectors in which these anti-emission facilities are installed, thus changing the rate of private income and finally the private consumer demand.

The model was conceived in the form of a linear system, that is, as a socalled "linear world" model. As long as computational conditions for numerical evaluation of extensive models are fulfilled only within the framework of linear structures, it will be inevitable to formulate such models on the basis of restricting assumptions regarding linear production- and demand implications.

Last but not least, it is necessary to point out the considerable statistical-empirical problems of data gathering and data processing that precede the numerical determination of parameters and coefficients of the total model. In principal these tasks can only be observed by major scientific teams and using the help of official statistical institutions. The individual scientist, as a rule, can only conceive "empty boxes" depicting, as exactly as possible, the technological and behavioral structures reflected by economic reality.

Footnotes

1 This means especially that the problem of quantity rates occurring in empirical models, is solved on the basis of value rates by formulating all variables in socalled "Leontief Units" (see Hatanaka, 1960).

2 "Scarce" raw materials are, hence, for instance petroleum or certain metals and ores in the FRG, whereas lime, cement, bricks, or similar materials are of the "non-scarce" type, so that production sectors like "Bau, Steine, Erden", or "Kohlegewinnung", remain components of the domestic economic production structure.

References

Chenery, H.B., 1956, Interregional and International Input-Output Analysis, in Barna, T. (Ed.), The Structural Interdependence of the Economy, New York.

Danzig, G.B., 1955, Upper Bounds, Secondary Constraints, and Block Triangularity in Linear Programming, Econometrica, 23.

Dorfman, R., Samuelson, P.A. and Solow, M., 1958, Linear Programming and Economic Analysis, New York.

Hatanaka, M., 1960, The Workability of Input-Output-Analysis, Ludwigshafen a.Rh.

Leontief, W.W., 1951, The Structure of the American Economy, 1919-1938, New York.

Meadows, D., 1972, Die Grenzen des Wachstums, Stuttgart.

Moses, L.N., 1955, The Stability of Interregional Trading Patterns and Input-Output Analysis, The American Economic Review, 45.

Rembold, G., 1975, Ein interregionales, intersektorales Wachstumsmodell auf der Basis der Input-Output-Analyse, in Funck, R.(Ed.), Karlsruher Beiträge zur Wirtschaftspolitik und Wirtschaftsforschung, Chapter 4.

Schumann, J., 1968, Input-Output-Analyse, Berlin.

Solow, R.M., 1959, Competitive Valuation in a Dynamic Input-Output System, Econometrica, 27.

Thoss, R., 1973, Ein integriertes Optimierungsmodell für die Planung des Umweltschutzes, in Menke-Glückert, P. et al., Planung für den Schutz der Umwelt, Materialien zum Siedlungs- und Wohnungswesen und zur Raumplanung, 2, Münster.

INDUSTRIAL INVESTMENT MODEL IN AN UNDERDEVELOPED AREA

P. Migliarese and P.C. Palermo
Politecnico di Milano
Milano - Italy

I. Introduction

This paper presents two mathematical models for the selection and location of industrial investment projects in an underdeveloped area.

The general characteristics of the problem are the following :

a) Productive investments are necessary in order to eliminate the emigration exodus from the area end to reduce the subemployment quota. Only industrial investments seem to guarantee, in the short run, a sufficient income and employment increase. Different criteria can be followed to find possible investment sectors : the degree of utilization of internal resources, the consistency with the external industrial structure, the capability to generate induced effects within the sub-region. In general, it is possible to define some alternative development strategies.

b) Due to the scarcity of available resources (investment, capital) it is reasonable to consider economic activities with high employment/capital ratio and to assume the minimization of investment costs as the objective function, guaranteeing both the desired employment level and productive levels, that are efficient from the economic point of view. Then, for each development strategy, which can be defined by a set of economically acceptable investment projects, the investment choice model selects the projects to be realized and the respective productive levels.

c) It is reasonable to assume that location costs of each activity sector in the sub-region do not affect significantly the previous economic evaluation.

The difference in the location costs between different infra-structured areas within the sub-region can be disregarded in comparison with investment costs. Choice of location is important with respect to the external effects : commuting or internal migration within the sub-region, environmental effects, undesirable effects on other non industrial activities (external diseconomies).

The location model assigns the investment projects selected from the former model to different alternative locations in the sub-region, minimizing total infrastructure and location costs and satisfying some constraints which aim at eliminating external diseconomies.

Each development strategy can be examined with the aid of these two hierarchically interconnected models. Investment and location costs (which must be supported by the private sector), infrastructure costs (which must be supported by the public sector) and external effects can be summarized in a balance-sheet, which is useful for the comparison and the comprehensive evaluation of different strategies.

II. An Investment Choice Model

1. General Framework

This model is applied to a set of productive investment projects and yields the following results : a) the selection of investment sectors and b) the specification of productive levels satisfying revenue, plant-size and employment constraints.

Assumptions on the exogenous variables : i) The problem of specifying possible productive sectors remains outside of the concern of this model. The model assumes that the possible industrial interventions, among which one must choose, are know. ii) For each industrial investment a production level range is defined by which the possibility of finding entrance markets within a reasonably accessible area is guaranteed.

Assumptions about productive investments :
The possible investments are defined in terms of the capital product ($C_i = C_i(P_i)$) and employment product ($L_i = L_i(P_i)$) functions and are characterized by scale and some type of agglomeration economies : i) the economies of scale are taken into account by assuming :

$$C_i(P_i) = k_i\, P_i^{\delta_i} \qquad 0 < \delta_i < 1 \qquad (1)$$

$$L_i(P_i) = h_i\, P_i^{\lambda_i} \qquad 0 < \lambda_i < 1 \qquad (2)$$

No substitutability between capital and labour factors at fixed production levels has been hypothesized. ii) the economies of agglomeration have been associated with the advantage accrueing to each industrial sector when the supply of raw and intermediate materials within the sub-region is possible and have reference to the consumer's transportation cost reduction. The price structure, in relation to the short and middle run period of the plan, is assumed to be known, at least in a relative sense. In particular, in the case of acquisition within the sub-region, the transportation cost has been assumed to be zero due to small spatial dimensions of the sub-region under study. In the case of supplies coming from outside the sub-region, the cost has been assumed equal to t_j, independently of the location of the consumption sector i.

For each investment the production scale presents minimal (P_i^t) and maximal (P_i^M) levels. In specifying this parameters, many different considerations occur : technological indivisibilities, necessity of guaranteing market absorption to the productive investment, diversification of the production, necessity of safeguarding environmental characteristics that are typical for this sub-region.

2. Structure of the model

As noted in the introduction, the aim of government intervention in the industrial sector is the creation of employment opportunities within the sub-region, opportunities that are capable of absorbing the structural emigration from the primary sector. In addition, such opportunities contribute to the reduction of the substantial income disequilibrium with respect to the 'visible' external zones. The evident scarcity of financial resources available through public channels or through private sources, leads us to assume the minimization of investment costs associated with the programmed industrial development as the objective of the model. In addition, the model, taking account of the short period of the plan, has a static character and refers to a single period.

The specific objective function assumed is as follows :

$$\min \sum_i p_i^* C_i(P_i) \qquad (3)$$

in which the parameter p_i^* represents the acquisition cost of the capital C_i derived from the typology of industrial settlement.

The constraints introduced are as follows :

(i) <u>revenue constraint</u> : revenue constraint was imposed for each project, at an exogenous fixed level, for the purpose of guaranteing the economic feasibility and the stability of the settlements, at least over a certain period of time. The agglomeration economies are involved in this constraint when the industries supplying the sector under study with raw or intermediate material are contemporaneously present, at the congruent production level, in the sub-region;

(ii) <u>employment constraint</u> : the employment objective associated with industrial development is globally imposed on the whole set of investments selected. The local income produced by employment generated in this way is functional for the objective regarding the reduction of the income disequilibrium relative to external areas.

The notations used are as follows :

$C_i = k_i P_i^{\delta_i}$: capital product function in industrial sector i;

$L_i = h_i P_i^{\lambda_i}$: employment-product function in industrial sector i;

$P_i =$ production level of sector i;

$p_i^\star =$ unit price of capital in industrial sector i;

$p_i =$ unit price of product i;

$w_i =$ unit wage in industrial sector i;

$\eta_i =$ depreciation of the plants in industrial sector during a single period

$\pi_i =$ unit revenue assumed for industrial sector i;

$J_i =$ set of supplying sectors to industrial sector i;

$N =$ employment to be generated;

$t_j =$ unit cost of transportation for external sector j;

x_i = binary selection variable associated with sector i;

a_{ij} = technical coefficient (input-output coefficient) j.

Through obvious analytical steps, the model can be formulated as follows :

$$\min C = \sum_{i=1}^{n} p_i^* C_i(P_i) x_i = \sum_{i=1}^{n} k_i' P_i^{\delta_i} x_i \quad (4)$$

$$\{p_i P_i - w_i' P_i^{\lambda_i} - (\eta_i' + \pi_i') P_i^{\delta_i} - \sum_{j \epsilon J_i} (p_j + (1-x_j) t_j) a_{ji} P_i\} x_i \geq 0$$

$$i = 1, \ldots, n \quad (5)$$

$$\sum_{i=1}^{n} h_i P_i^{\lambda_i} \cdot x_i \geq N \quad (6)$$

$$P_i^t \leq P_i \leq P_i^M \quad i = 1, \ldots, n \quad (7)$$

$$x_i = 0, 1 \quad (8)$$

The proposed model is non linear in the P_i variables and in the decisional binary variables x_i. The difficulty of formulation has lead us to specify an heuristic algorithm for solution that, taking advantage of the structure of the problem, enables us to obtain a 'good' choice between different alternatives with limited effort. The possibilities presented by the agglomerative economies are specified before the application of the solution procedure and enable the construction of 'agglomerative projects', including some interdependent sectors. In particular, each agglomerative project is composed of (a) a sector producing (from the point of view of the sub-region) final goods; (b) one or more sectors supplying the preceding sectors with intermediate goods.

Each agglomerative project is consistent in the sense that the supplying sectors have productive scales that are connected with the productive scale of the final sector, on the basis of the sectoral interdependence (the technical coefficients are assumed known). In this way, it is possible to elaborate unique capital-product and employment-product functions for the entire agglomerative project.

Different alternatives of development strategies are assumed for the sub-region and the methodology presented here can be applied to each of these. In this way, within a selected strategy, one or more agglomerative projects that can realize such strategy can be constructed. Then, the solution scheme operates a selection within a set of projects containing two different types of projects : (a) a set of independent projects (b) one or more agglomerative projects (corresponding to a defined development strategy).

If we want to consider the case in which a project can be selected either within an agllomerative project or as an independent one, we must introduce some 'incompatibility' constraints between different projects. This can be accomplished by means of the definition of a set S_i which contains all the projects that are incompatible with the presence of project i in the solution. The procedure defined later takes into account this opportunity. The proposed algorithm consists in the following phases :

(i) <u>Projects ordering, satisfaction of revenue and incompatibility constraints</u> : since the left side of the constraint (5) is a monotonic function of P_i, it is possible to determine a level P_i^{π}, below which the revenue constraint relative to project i is not satisfied. Then, if $P_i^{\pi} > P_i^{M}$ the project i is excluded from further considerations. Define :

$$P_i^m = \max \{P_i^t, P_i^{\pi}\} \qquad (9)$$

and
$$\rho_i(P_i) = \frac{C_i(P_i)}{L_i(P_i)} = \frac{k_i}{h_i} P_i \quad (\partial_i > \lambda_i) \qquad (10)$$

the capital/labour ratio. This function will increase or decrease according to whether $\delta_i > \lambda_i$ or $\delta_i < \lambda_i$ respectively. For each investment project i, let $\rho_i(\overline{P}_i)$ be the minimal value of the capital/labour ratio. The different projects i∈I are then reordered on the basis of increasing values of $\rho_i(\overline{P}_i)$. The procedure initializes with the first element in this list. At the k-th stage, a project k is selected from the list; project k and all project k'∈S_k are eliminated from the list, thus obtaining the up-dated list. In this way the incompatible constraint is respected. In order to have an exhaustive enumeration, it is possible to apply this procedure iteratively.

(ii) <u>Exploration phase at stage k</u> :

Let be $\rho_k(\overline{P}_k)$ the ratio considered at stage k. In addition to project k, consider the sets of projects considered at previous steps defined as follows. At step i :

$$(\delta_i - \lambda_i) > 0, \; \rho_i(P_i^m) \leq \rho_k(\overline{P}_k); \; \rho_i(P_i^M) > \rho_{k+1}(\overline{P}_{k+1}) \quad (11)$$

$$i = 1,\ldots,n_k;$$

Let be $P_i^\star$ the production level for which

$$\rho_i(P_i^\star) = \rho_{k+1}(\overline{P}_{k+1}) \text{ and } L_i(P_i^\star) \text{ is the corresponding employment}$$

level. At stage j :

$$(\delta_j - \lambda_j) > 0, \; \rho_j(P_j^m) \leq \rho_k(\overline{P}_k), \; \rho_j(P_j^M) \leq \rho_{k+1}(\overline{P}_{k+1}) \quad (12)$$

$$j = 1,\ldots,m_k$$

k+1 : denotes the first element in the up-dated list.

We define the projects specified in this way as <u>active</u> projects. In addition, projects carried out at previous stages to the maximum production level are defined as <u>completed</u>.

Let be N_k the total employment generated by completed investments.

(iii) <u>Selection phase at stage k</u>

The following test is accomplished at every stage :

$$\sum_{i=1}^{n_k} L_i(P_i^\star) + \sum_{j=1}^{m_k} L_j(P_j^M) + L_{k+1}(\overline{P}_{k+1}) \leq \Delta N_k \quad (13)$$

where $\Delta N_k = (N-N_k)$ represents the employment that still must be generated at stage k.

If such a relationship is satisfied, all the projects thus considered are selected at the following production levels:

i : at production level P_i^*

j : at production level P_j^M

k+1 : at production level $\overline{P}_{k+1}$

and the procedure continues to the next stage (k+1).

If, instead, such a relationship is not satisfied, the following optimization sub-problem will be considered (for convenience, let A^k be the set of active projects at stage k):

$$\min \sum_{i \in A^k} C_i(\tilde{P}_i + y_i) \qquad (14)$$

$$\sum_{i \in A^k} L_i(\tilde{P}_i + y_i) \geq \Delta N_k \qquad (15)$$

$$y_i \geq 0 \quad ; \quad i \in A^k, \qquad (16)$$

where $\tilde{P}_i$ is the production scale already assigned to the preceding stage or the threshold level $\overline{P}_k$ of the investment corresponding to the current stage.

With respect to the primary problem, a problem which generally presents a notably smaller number of variables is obtained. In addition, the reduced employment quota to be allocated justifies a linearization of functions $C_i(P_i)$ and $L_i(P_i)$ around the $\tilde{P}_i$ values, for active projects only.

III. A Location Model

1. Structure of the model

The previous model determines a set of efficient projects. Let I = {1,...,n} be now the set of selected projects. The 'public interest' is taken into account in the industrial employment constraint and in the choice of objective function. Now the analysis must be detailed to consider the social costs and the external effects, related to the location of economic activities: infrastructure costs of the area destinated to industrial use; environmental constraints and constraints designed to rule out incompatible locations of different activities.

The location model also considers the private costs specifically connected with the location of each project in a determined site. Areas of different amplitude can be equipped for industrial use in different sites. Let J = {j = 1,...,m} be the set of possible locations and

T_{jk}, ($k=1,\ldots,K(i)$), $T_{jk+1} > T_{jk}$ the set of possible amplitudes, according to urbanistic constraints, of the industrial area in the location j. Let $f_j(T_{jk})$ be the corresponding cost of infrastructuralization, at a standard service level, while $\{y_{jk}\}$ is the set of decision variables : $y_{jk} = 1$ means that an industrial area of amplitude k will be equipped in j; $y_{jk} = 0$ otherwise; obviously

$$\sum_{k=1}^{K(j)} y_{jk} \leq 1 \quad , \quad \forall j \in J \qquad (17)$$

Some locations can be mutually incompatible e.g. the industrial development of an area excludes the one of a neighbor area either for resource distribution reasons or for external diseconomies. Hence, the constraints can be set as follows :

$$\sum_{l \in ME_j} \sum_{k=1}^{K(j)} y_{lk} + \sum_{k=1}^{K(j)} y_{jk} \quad \forall j \in J \qquad (18)$$

where $ME_j \subset J$ is the set of locations mutually incompatible with j and with each other.

If project i is located in j, location costs c_{ij} must be supported by the enterprise; let x_{ij} be the set of decision variables of project assignment to alternative locations. The external effects can exclude in some locations some projects :

$$\sum_{i \in E_j} x_{ij} = 0 \quad , \quad \forall j \in J$$

where $E_j \subset I$ is the set of projects which cannot be located in j (e.g. due to pollution, noise, etc.).

Moreover the simultaneous location of different activities in the same area can produce external diseconomies; then

$$\sum_{i \in D_j} x_{ij} \leq 1 \quad , \quad j \in J \quad ,$$

where $D_j \subset I$ is the set of activities which are mutually exclusive in j. Finally the location of the projects in the same area must respect physical constraints :

$$\sum_{i=1}^{n} t_i x_{ij} \leq \sum_{k=1}^{K(j)} T_{jk} y_{jk},$$

where t_i is the spatial requirement of the activity i. Thus the following location model (integer programming model) can be constructed :

$$\min_{(x_{ij},y_{jk})} \sum_{j=1}^{m} \sum_{k=1}^{K(j)} f_j(T_{jk}) y_{jk} + \sum_{i=1}^{n} \sum_{j=1}^{m} c_{ij} x_{ij} \quad (19)$$

$$\sum_{k=1}^{K(j)} y_{jk} \leq 1 \qquad j \in J \quad (20)$$

$$\sum_{l \in ME_j} \sum_{k=1}^{K(j)} y_{lk} + \sum_{k=1}^{K(j)} y_{jk} \leq 1 \qquad j \in J \quad (21)$$

$$\sum_{j=1}^{m} x_{ij} = 1 \qquad i \in I \quad (22)$$

$$\sum_{i \in E_j} x_{ij} = 0 \qquad j \in J \quad (23)$$

$$\sum_{i \in D_j} x_{ij} \leq 1 \qquad j \in J \quad (24)$$

$$\sum_{i=1}^{n} t_i x_{ij} \leq \sum_{k=1}^{K(j)} T_{jk} y_{jk} \qquad j \in J \quad (25)$$

$$x_{ij} = 0,1; \; i \in I, \; j \in J \quad (26)$$

$$y_{jk} = 0,1; \; j \in J; \; (k=1,\ldots,K(j)) \quad (27)$$

2. An Algorithm

The algorithm is based on implicit enumeration techniques. Feasible solutions $\{y_{jk}\}$ are generated with an efficient procedure satisfying (20), (21), (27). The activity allocation is made by the solution of a modified "special transportation program". Actually, the objective function in the allocation phase is

$$\min_{x_{ij}} \sum_{i=1}^{n} \sum_{j \in \bar{J}} c_{ij} x_{ij}, \text{ where } \bar{J} = \{j | \sum_{k=1}^{K(j)} y_{jk} = 1\} ; \quad (28)$$

(22), (25) and (26) are the classical special transportation program constraints; while (23) and (24) are the additional constraints introduced in the model. A typical difficulty of the model derives from the alternative area amplitudes in each location. The problem has been solved by generating in a first phase a feasible choice of location at maximum amplitude level. It is then attempted to improve the solution by reducing in a systematic way the amplitude of the areas. Compatibility and feasibility tests which are derived from the constraints and lower bounds on the variables allow "to cut" the enumeration. The dimensions of the sets I and J facilitate the computational effort in order to determine the optimal solution of the model. The essential characteristics of the algorithm are the following :

a) Enumeration on the main tree

(i) <u>Initial step</u>

$$y_{jK(j)} = 1 \quad \forall_j \in J \qquad (29)$$

Compatibility test :

$$\text{if } \gamma_j = \sum_{l \in ME_j} y_{1K(l)} + y_{1K(j)} > 1, \; j \in J \qquad (30)$$

branch ("close" a location $j \in J$, which is selected according to a heuristic criterion); if $\gamma_j = 1$, solve the allocation problem (see c)). Let $\bar{z}$ be the value of the objective function : $z^o = \bar{z}$ is the value of the best solution obtained so far.
Go to b).

(ii) <u>Branch step at level h</u> (h=1,...,m)

$$\text{Let } J_1^{h-1} = \{j | y_{jK(j)} = 1\} \; ; \sum_{j \in J_1^{h-1}} y_{jK(j)} = m-h+1 \qquad (31)$$

Branch : select $j^* \in J_1^{h-1}$ and set

$$y_{j^*K(j^*)} = 0; \; \{J_1^h = J_1^{h-1} - j^*\}. \qquad (32)$$

Compatibility test :

$$\text{if } \gamma_j = \sum_{l \in ME_j} y_{1K(l)} + y_{jK(j)} > 1, \; j \in J_1^h \qquad (33)$$

go to the branch step at level h+1; if $\gamma_j = 1, \forall_j \in J_1^h$, check the feasibility test :

$$\sum_{j \in J_1^h} T_{jK(j)} y_{jK(j)} \geq \sum_{i=1}^n t_i \qquad (34)$$

If the test is not satisfied backtrack ("open" again j^* and return to the level h-1). Otherwise, solve the allocation problem. Let z^h be the value of the objective function. If $z^h < z^o$, set $z^o = z^h$. Go to b).

b) Enumeration on the secondary three

(i) <u>Initial step</u>

$$y_{jK(j)} = 1, \quad \forall_j \in J_1^h; \quad z^o. \tag{35}$$

(ii) <u>Branch step at level s</u> (s = 1,...,m^h)

$$m^h = \sum_{j \in J_1^h} y_{jK(j)} \tag{36}$$

Let

$$J_1^{s-1} = \{j | y_{jK(j)} = 1\} \;;\; J_2^{s-1} = \{J_1^h - J_1^{s-1}\};$$

$$\sum_{j \in J_1^{s-1}} y_{jK(j)} = m^h - s + 1. \tag{37}$$

Choose $j^* \in J_1^{s-1}$ and consider the alternatives $y_{j^*k} = 1, (k=1,...K(j^*)-1)$. Set $J_1^s = \{J_1^{s-1} - j^*\} \;;\; J_2^s = \{J_2^{s-1} + j^*\}$. (38)

Feasibility test : if

$$\sum_{i=1}^{n} t_i > \sum_{j \in J_1^s} T_{jK(j)} + T_{j^*\bar{k}} + \sum_{j \in J_2^{s-1}} \sum_{k=1}^{K(j)-1} T_{jk} y_{jk} \tag{39}$$

then $y_{j^*k} = 0, \quad \forall k=1,...,\bar{k}$.

If $\bar{k} = K(j^*)-1$ backtrack; if $\bar{k} < K(j^*)-1$ compute the lower bound $LB^s(j^*k)$, $k > \bar{k}$ (see point b.iii)). Reorder the lower bounds in a decreasing list. If $LB^s(j^*k) > z^o, \forall k > \bar{k}$, backtrack; otherwise solve the allocation problem on the terminal nodes such that $LB^s(j^*k) < z^o$. Let z^s be the value of the objective function. If $z^s < z^o$ set $z^o = z^s$. If the node is not a terminal one branch : $y_{j^*k^*} = 1$ where j^*k^* is the alternative with minimum lower bound; go to the branch step at level s+1. If the secondary tree derived from the solution $y_{jK(j)} = 1$ $\forall j \in J_1^h$, has been completely explored go to the branch step at level h+1 on the main tree.

(iii) <u>Lower bound</u>
Two components are summed up : infrastructure costs (Ic) and location

costs (LC).

$$IC^s(j^*k) = \sum_{j \in J_2^{s-1}} \sum_{k=1}^{K(j)-1} f_j(T_{jk})y_{jk} + f_{j^*}(T_{j^*k}) + ICM(J_1^s) \quad (40)$$

where :

$$ICM(J_1^s) = \min_{\{\alpha jk\}} \sum_{j \in J_1^s} \sum_{k=1}^{K(j)} f_j(T_{jk}) \alpha_{jk} \quad (41)$$

subject to the constraints :

$$\sum_{j \in J_1^s} \sum_{k=1}^{K(j)} T_{jk}\alpha_{jk} \geq \sum_{i=1}^{n} t_i - \sum_{j \in J_2^s} \sum_{k=1}^{K(j)} T_{jk}y_{jk}; \quad (42)$$

$$\sum_{k=1}^{K(j)} \alpha_{jk} \leq 1 \quad , \quad \forall j \in J_1^s; \quad (43)$$

$$\alpha_{jk} = 0,1. \quad (44)$$

The model can be solved via dynamic programming methods. A parametric procedure can be developed since the algorithm must be repeated many times, different amplitudes being assigned to the areas.

$$LC^s(j^*k) = \min_{\{x_{ij}\}} \sum_{i=1}^{n} \sum_{j \in J_1^h} c_{ij}x_{ij} \quad \text{s.t.} \quad (45)$$

$$\sum_{i=1}^{n} t_i x_{ij} \leq \sum_{k=1}^{K(j)-1} T_{jk}y_{jk} \quad j \in J_2^{s-1} \quad (46)$$

$$\sum_{i=1}^{n} t_i x_{ij} \leq T_{jK(j)} \quad j \in J_1^s \quad (47)$$

$$\sum_{j \in J_1^h} x_{ij} = 1 \quad i \in I \quad (48)$$

$$\sum_{i \in E_j} x_{ij} = 0 \quad j \in J_1^h \quad (49)$$

$$x_{ij} = 0,1. \quad (50)$$

This model belongs to the class of "special transportation models". One can take into account constraint (49) by setting $c_{ij} = \infty$, $\forall i \in E_j$, $j \in J_1^h$.

c) The Allocation problem

The Srinivasan-Thompson (1973) algorithm is adopted in order to solve the special transportation problem. The algorithm is based on a branch and bound procedure and utilizes the standard transportation routine as sub-problem. Its solution is not feasible for the allocation poblem if the constraints $x_{ij} = 0,1$, $i \in I$, are not satisfied. Consider $\bar{J}$ such that $0 < x_{ij} < 1$ ($\bar{J}$ heuristically determined). The set of solutions can be partitioned into two subsets :

1) the one where $x_{i\bar{j}} = 1$, 2) the other where $x_{i\bar{j}} = 0$. Lower bounds on the optimal solution are obtained by solving the transportation problem respectively under constraint (46) and constraint (47). It is easy to take into account constraint (49) : if $x_{i\bar{j}} = 1$, then set $c_{1\bar{j}} = \infty$, $\forall_1 \in D_j$, $1 \neq i$. Hence constraint (49) reduces the computational effort to obtain the solution to the special transportation problem.

References

De Maio, A., Migliarese, P., Palermo, P.C., 1974, "A subregional planning model", in Proceedings of the 2th Polish-Italian Conference on "Application of systems theory to economy management and technology", Pugnochiuso, Italy.

Scrinivasan, V., Thompson, G.L., 1973, "An algorithm for assigning uses to sources in a special class of transportation problems", Operations Research, 21, no. 1.

COST-BENEFIT ANALYSIS AND OPTIMAL CONTROL THEORY FOR
ENVIRONMENTAL DECISIONS : A CASE STUDY OF THE DOLLARD ESTUARY

P. Nijkamp and C. Verhage
Free University, Amsterdam
The Netherlands

I. Introduction

Already since Dupuit's pioneering study, "On the Measurement of the Utility of Public Works" (1844), economists try to provide guidelines to policy-makers when they are in the awkward position of selecting one public investment project out of several alternatives. The reputation of great economists (Pigou, Keynes, Wicksell, Samuelson, e.g.) is related to this problem of project evaluation. Yet there is hardly any practicable solution found. Therefore it is still and attractive field of study.

This paper concentrates on optimal decision rules for regional development alternatives with environmental repercussions. After a discussion of environmental impact studies and multiple-criteria decision problems, the use of optimal control theory is proposed as an expedient to deal with dynamic development problems.

The various evaluation problems are analysed within the aforementioned framework of a dynamic control model, in which a modified cost-benefit analysis plays a major role.

Given a formal specification of the economic-ecological decision framework at hand, a set of optimal decision rules is derived.

Next, the foregoing theoretical model is applied to a regional development program in the Northern part of the Netherlands (viz. the Dollard), where a friction between development efforts in the field of water resources and environmental preservation became apparent.

The public investment project investigated is a <u>drainage canal annex new sea-dike before the coast in the Dollard-area</u>. The alternative decisions to be considered by the Minister of Public Works differ basically in the treatment of a unique birds-reservation : preserving or sacrificing. Contrary to what might be expected the policy-maker did not appoint a committee of experts in order to investigate what were the best plan, so that he could base his decision upon their advice. The statesman decided recently more or less intuitively in favour of the birds-reservation, without knowing quite well what the monetary advantages and disadvantages would be.

We found this procedure remarkably enough to inspect the possibilities of determining the policy-maker's implicit evaluation of the natural area based on a comparison of the monetary consequences of the various plans. The method is in fact a micro-analytic version of earlier research about implicit social preference functions (Nijkamp and Somermeyer, 1971; Nijkamp and Paelinck, 1973, and Nijkamp, 1975). The basic assumption is that the decision actually taken reveals the policy-maker's preference. From the revealed preference can be inferred that the decision-taker evaluated the chosen alternative higher than the rejected alternative(s). If this would not be the case, i.e., if the rejected plans were preferred more than the chosen one, it is said that the choice behaviour is inconsistent, which is ruled out in the theory of revealed preference. With respect to our project the revealed preference approach implies that the Minister's choice of the alternative that saved the birds-reservation, should have such a value that the additional costs compared with the next-best alternative are fully justified. See for other, more macro-economic oriented, uses of revealed preference concepts a.o. Friedlaender (1973), Paelinck (1973) and Rausser and Freebairn (1974).

Some empirical aspects of the Dollard-project will be presented in this paper according to the lines set out above. Finally an evaluation of the method employed and an outline for future research will be given.

II. Impact analysis

The wider the sphere of influence of any decision, the more it will cause unintended effects, which may be advantageous or disadvantageous depending on the degree of external spill-overs. This is equally valid for a public investment project.

All the intended and unintended consequences of a certain decision can be considered as its impact vector $\underline{r}$ with typical element

$r_j (j=1,\ldots,J)$. The impacts are only technical and physical effects of the project concerned; they are not yet transformed by means of a particular price vector in economic evaluations.

As the actual plan implementation has been chosen as the best plan out of a set of N alternatives there are N impact vectors $\underline{r}_n$, $(n=1,\ldots,N)$, which form together the following impact matrix :

$$\underline{R} = \{\underline{r}_1,\ldots,\underline{r}_N\} \tag{II.1}$$

Matrix $\underline{R}$ is of order JxN, since there are J types of impacts due to each of the N plan alternatives; see Nijkamp (1974b).

In the last decade we have learnt by bitter experience that the environmental consequences of public (and private) investments should appear in the impact matrix. In digging a drainage canal annex sea-dike these impacts consist of e.g. the number of hectares of aquatic area that is lost, the portion of the coastal wetlands of an estuary that is converted to terrestrial domain, the removal of a unique bird-nesting area, etc. It requires a painstaking inventory of the natural environment that is threatened by the implementation of the investment project. The voluntary work of an active local environmental study group can be of great importance, because otherwise a huge data bank would be needed with information on the environmental qualities of each hectare of the national territory. The more accurate the ecological impacts are described, the greater the chance that the natural damage will be taken into account more accurately.

In addition to the environmental impacts a public investment project like the drainage canal, causes also impacts of a hydrological, physical planning and economic nature.

The <u>hydrological</u> effects are :

- drainage of raining water from the polders;
- discharge of purified sewage;
- navigation of barges;
- protection against water floods.

The consequences in the <u>physical planning</u> framework are :

- changes in the pattern of land occupation, i.e. more land for

recreational and residential or commercial purposes and less for agricultural use;
- improvement of infrastructural communications, either by road or by waterway.

The economic impacts consist inter alia of :
- direct influences on the agricultural output due to changes in the amount of arable land;
- direct effects on output, employment and income in other sectors owing to the regional multiplier of the investments concerned;
- indirect changes of activities in the particular region because its location qualities with respect to other locations are improved.

This list of impacts has only an illustrative purpose and is not exhaustive. The impacts of the dike and canal project are related to the conversion of the present spatial lay-out of that area into a different lay-out. To every change of lay-out belongs a specific combination of impacts. It are the civil-engineering works of the investment project and the land occupation they require which are responsible for the changes in land use and the related impacts.

When the analysis is operationalized in section 4, we will represent the impacts by the alterations in land use to which the several plan alternatives give rise. The impact-vector, $\underline{r}_n$, of the n-th alternative will then contain the percentage distribution of the available land over various purposes, r_{jn}, (j=1,...,J; n=1,...,N). In as far as the multi-purpose use of a unit of land or water (in general, of space) is concealed by employing only land use variables, all the impacts of a particular use have to be added together. This will be set out in more detail in the next sections.

III. Evaluation of Projects with Multiple Criteria

Traditionally, the analysis of public projects has been based upon the concept of consumers' surplus on which an early essay was published by Dupuit in 1844. Despite improvements of the concept by Marshall and Hicks it was not feasible for practical applications.

In an attempt to evade the aggregation of individual welfare gains and losses some authors developed as a substitute for this type of social welfare function the policy maker's objective function (Little, 1952 and Bergson, 1954). Such an objective function contains the personal preferences of the political authority and it is assumed that in one way or another his judgements reflect the preferences of his voters. In this construction there is not necessarily a conflict with Arrow's Impossibility Theorem which would arise if a social welfare function should represent exclusively individual preferences (Arrow, 1963). By means of the decision-maker's objective function the social utility is maximised under the side condition of the available budget.

In empirical situations, however, cost-benefit analysis and cost-effectiveness analysis are used to evaluate concrete projects. The main difference in both approaches is the way the wide diversity of impacts is treated.

In cost-benefit analysis the multiplicity of dimensions are transformed into one common denominator by expressing all impacts in monetary units. In terms of section II : the physical effects, r_{jn}, of matrix $\underline{R}$ are multiplied by a set of corresponding prices or rates of substitution p_j, to obtain the matrix of monetary evaluations $\underline{M} = \{m_{jn}\}$, $(j=1,\ldots,J; n=1,\ldots,N)$.

$$\hat{\underline{p}} \cdot \underline{R} = \underline{M} \qquad\qquad\qquad (III.1)$$

where $\hat{\underline{p}}$ is a diagonal matrix of order J, with the prices p_j as elements on the main diagonal. Only those impacts which have no definite generally accepted economic value, like human life, natural beauty or other intangibles are mentioned as P.M. items. The relevant prices of these imponderables in matrix $\hat{\underline{p}}$ might be ∞. Very often the importance of these immaterial qualities was neglected in the evaluation process, since the other impacts were aggregated into a number of dollars or guilders. That is the reason why cost-benefit analysis looks so suspicious to anybody. We think, however, that an intelligent user of cost-benefit analysis, who gives credit to the P.M. recorded impacts, may take advantage of it. There are several drawbacks the user should be aware of : the lifetime of any project is hard to estimate, the rate of discount is an arbitrary number, and efficiency considerations dominate equity considerations.

Cost-effectiveness analysis tries to leave the multi-dimensional character of the various impacts unaffected. The analysis studies the way in which a predetermined set of goals is attained by the various plan alternatives. The alternative which causes the highest total effectiveness is selected for implementation.

Now the researcher is freed from the task to transform the dimensions into one common measure, but the decision-maker himself has to compare anyhow the performances on the different goals of the plan alternatives. Hence the same problems exist in cost-effectiveness as in cost-benefit analysis; only the determination of the trade-offs is postponed to the confrontation with the personal value judgements of the politician in charge. He has to decide, for instance, whether an improvement of the relative inequality of income of region X with respect to the national average with 1 % is equally (or less, more) important as (than) a decrease in GNP of 0,5 %, assuming that the two plan alternatives concerned have the same achievement on all other goals, and cost the same amount of money.

The selection problem of cost-effectiveness analysis can be described as selecting that alternative that attains the highest scores on the goals with a given budget. Instead we could formulate : <u>selecting that alternative that is cheapest and meets all the predetermined minimal scores on the different goals</u>. According to Simon (1958) we call the minimal scores on the goals the <u>aspiration</u> levels, which form the <u>aspiration vector</u> $\underline{r}^s$, of order J.

Such a minimizing problem is generally formalised as :

$$\min \eta = \sum_{n=1}^{N} f_n(\underline{r}_n) \cdot \delta_n$$

subject to $\quad \underline{r}_n \geqslant \underline{r}_n^s \cdot \delta_n$

$$\sum_{n=1}^{N} \delta_n = 1 \qquad\qquad\qquad\text{(III.2)}$$

$$\delta_n = 0,1$$

$$\underline{r}_n \geqslant \underline{0}$$

where : f_n = the costs of alternative n which gives rise to the impact vector $\underline{r}_n$;

δ_n = a zero - one valued variable that restricts the optimal program to one and only one alternative.

The dual of the cost minimizing problem (III.2) is the maximizing problem of the imputed value of the aspiration levels, given side conditions on costs.

$$\max \omega = \sum_{n=1}^{N} \underline{\pi}'_n \underline{r}^*_n \delta_n$$

subject to

$$\underline{\pi}_n \leq \left(\frac{\Delta f_n}{\Delta \underline{r}_n}\right) \cdot n$$

$$\sum_{n=1}^{N} \delta_n = 1 \qquad (III.3)$$

$$\delta_n = 0, 1$$

$$\underline{\pi}_n \geq \underline{0}$$

where $\underline{r}^*_n$ = the optimal solution from (III.2);

$\underline{\pi}_n$ = the J-ordered vector of (imputed) shadow prices associated with the active constraints of the aspiration levels of $\underline{r}^s$;

$\frac{\Delta f_n}{\Delta \underline{r}_n}$ = the marginal change of the cost function in (III.2), given a discrete change in $\underline{r}_n$

Program (III.3) attempts to <u>maximise the imputed value of the project impacts, given a set of side conditions that guarantees an efficient allocation of available funds</u>. It is obvious that such a program bears a close resemblance to a cost-benefit approach of investment projects. In the remainder of this section we will derive an expression like (III.3) for the selection problem of cost-benefit analysis, which expression should easily be adopted to our empirical study of the dike and canal project.

Assuming that we are able to gauge the whole price matrix, $\hat{\underline{p}}$, of (III.1) we obtain in the vector $\underline{m}_n$ the economic valuations of all impacts, both positive and negative, of alternative n (n=1,...,N).

Next, one has to deduct the negative amounts from the positive amounts. By denoting the resulting surplus per alternative plan by B_n, it represents the net social benefits as the difference between the direct and indirect returns and the operating costs and social losses.

Since we have to take into account that any investment project extends into the future we should adapt the determination of B_n for intertemporal aspects. We have to define for each period t of the whole project lifetime T an impact matrix $\underline{R}(t)$, a price matrix $\underline{\hat{p}}(t)$ and a plan surplus $B_n(t)$, (t=1,...,T).

The decision criterion, can now be specified as :

$$\max \omega = \sum_{t=1}^{T} \{B_n(t) - I_n(t)\} \delta_n \cdot (1+\rho)^{-t} \qquad (III.4)$$

or as a continuous system :

$$\max \omega = \int_0^T \{B_n(t) - I_n(t)\} \delta_n \cdot e^{-\rho t} \cdot dt$$

subject to $\sum_n \delta_n = 1$ \qquad (III.5)

$\delta_n = 0,1$

where $I_n(t)$ = the amount of investment in the plan alternative implemented at time t.

ρ = the social rate of discount

Expression (III.5) is the usual formulation of the continuous cost-benefit problem of an investment project, in which, however, it is explicitly stated that only one of the various plans should be adopted. Regional applications of cost-benefit analyses with environmental impacts are studied by Klaassen et al. (1974).

IV. An Optimal Control Model for Alternative Land Use Projects

In this section an attempt will be made to employ optimal control methods as a tool for analysing administrative decision-making in land use projects with environmental repercussions. Obviously, alternative projects are, in general, evaluated by means of trade-off schemes for various attributes of the investment problem in question (see section III). The assumption will be made here that the multiple objective criteria can be 'translated' into a common monetary 'language' so as to maximize the expected net social regional surplus of the public investment decisions at hand (for a generalization see section IX).

The project elements of our problem are mainly formed by the various hydrological works (like digging a canal and constructing a sluice). The impacts r_j (j=1,...,J) of this hydrological project are mainly determined by the spatial trajectory of the dike and canal project. These impacts include <u>inter alia</u> environmental repercussions, loss of agricultural areas, regional investment multiplier effects, and improved quality of other agricultural areas (see section VI). Evidently, the value of the effect vector $\underline{r}_n$ is determined by the location of the project concerned, so that the impact function is essentially provided with spatial co-ordinates.

For each project n (n=1,...,N) the attributes r_{jn} are formed by land use variables, viz. the size of an area earmarked for a certain purpose. In our study three different projects (i.e., N=3) will be distinguished each of which is characterized by three major attributes (i.e., J=3), viz. the amount of land use for hydrological, environmental and agricultural purposes. Other forms of land use (like residential, recreational and industrial uses) will be left out of consideration, since these uses are less relevant here and do not show any significant difference for each of the alternative projects. Obviously, the following additivity condition holds for the area in question :

$$\sum_{j=1}^{J} r_{jn} = \bar{r}_n, \forall n \qquad (IV.1)$$

where $\bar{r}_n$ is the known size of the area related to project n, and r_{jn} (j=1,...,J) the successive sizes of a sub-area earmarked for the jth type of land use. It is obvious that both the present state of land use as well as the land use pattern after a transformation should satisfy this constraint.

As indicated above, the provisional assumption will be made that the decisions concerning land use patterns associated with a certain project are guided by the criterion of a maximum total present value of net social returns from the area in question. These net social returns reflect the benefits minus the costs of the project concerned.

The benefits are according to Warnke, Terre and Ameiss (1973) composed of <u>direct</u> benefits (accruing from the direct output of the investment), <u>indirect</u> benefits (any other monetary benefit attributable to the investment) and <u>intangible</u> benefits (benefits not subject to monetary quantification). The costs reflect the direct capital outlays for constructing the project (i.e., for transforming a given land use pattern into a new one), the factor supply and overhead expenses, the opportunity costs and the social costs of undertaking the project in question.

Therefore, the decision criterion can be specified as:

$$\max \omega = \int_0^T \{B_n(t) - T_n(t)\} \delta_n e^{-\rho t} dt, \qquad (IV.2)$$

where $B_n(t)$ is a scalar-valued function representing the expected net social benefits of the nth project at time t, and $I_n(t)$ the investment costs for social overhead capital at time t in order to transform the area in question from the initial state to the new state associated with the nth project. Obviously, the 0-1 paramter δ_n satisfies the condition:

$$\sum_{n=1}^{N} \delta_n = 1$$
$$\delta_n = 0,1 \qquad (IV.3)$$

The arguments of $B_n(t)$ are ultimately formed by the vector $\underline{r}_n$, so that the net social benefits of plan n are a function of the land use pattern associated with this plan. It should be noted that $B_n(t)$ is not necessarily a separable benefit function, so that $B_n(t)$ is not <u>a priori</u> equal to $\sum_{j=1}^{J} B_{jn}(t)$, where B_{jn} represents the net benefit of the jth type of land use related to plan n. Therefore the normal expression of the social benefit function is:

$$B_n(t) = B_n(r_{1n}, \ldots, r_{Jn}) \qquad (IV.4)$$

It is clear that (IV.1) can be rewritten as:

$$r_{1n} = \bar{r}_n - r_{2n} - \ldots - r_{Jn}, \qquad (IV.5)$$

and, next, be substituted into (IV.4), so that (IV.4) can be written as a function of $r_{2n},\ldots,r_{Jn}$ only. Relationship (IV.5) indicates that r_{1n} can be calculated as a resultant of the other land use decisions.

Finally, land use policy and infrastructural design is, in general, an <u>irreversible</u> dynamic process. For example, a canal once constructed is hard to transform into an agricultural area. The degree of transforming effectively a piece of land from one use into another use depends obviously on the amount of <u>(social) overhead investments</u> $I_n(t)$. Assuming a <u>constant marginal capital output ratio for overhead investments</u>, one may specify the following differential equation:

$$\underline{\dot{r}}_n = \underline{\varphi}_n I_n \qquad (IV.6)$$

where a dot represents the time derivative of the vector of variables in question and $\underline{\varphi}_n$ a vector with (positive) marginal project effectiveness coefficients φ_n of overhead investments. Furthermore, the irreversibility constraint implies:

$$\varphi_{jn} I_n \geq 0 \qquad (IV.7)$$

irrespective of the direction of transformation of the successive land uses.

It should be noted that (IV.1) can also be rewritten in differential equation form as:

$$\dot{r}_{1n} = -\dot{r}_{2n} - \ldots - \dot{r}_{Jn}, \qquad (IV.8)$$

so that $\dot{r}_{1n}$ can also be considered as a remainder to be calculated once the other variables are known. This implies the number of decision variables is equal to J-1 (without altering the essential structure of the decision problems at hand).

Given the known initial state $\underline{r}_0$ of the system, the foregoing dynamic decision model can be synthesized as follows:

$$\max \omega = \int_o^T \{B_n(t) - I_n(t)\} \delta_n e^{-\rho t} dt$$

s.t.
$$\sum_{n=1}^N \delta_n = 1$$

$$\delta_n = 0,1 \qquad , \forall n$$

$$\underline{\dot{r}}_n = \underline{\varphi}_n I_n \delta_n \qquad , \forall n \qquad \text{(IV.9)}$$

$$\sum_{j=1}^J r_{jn} = \bar{r}_n \delta_n \qquad , \forall n$$

$$I_n \delta_n \geq 0 \qquad , \forall n$$

$$\underline{r}_{t=0} = \underline{r}_o$$

It is obvious that the previous model is essentially a 0-1 optimal control model.[1] The vector $\underline{r}_n$ is a vector of <u>state</u> variables, whereas δ_n and I_n represent <u>control</u> variables. The <u>system</u> (or <u>transition</u>) equations are represented here by the investment equation for land use dynamics. The previous 0-1 optimal control model is rather hard to solve, as there are hardly efficient solution algorithms for 0-1 optimal control problems. This implies that a definite conclusion concerning the optimality of a certain project can only be drawn by comparing all N individual projects mutually.

However, before proceeding with such an analysis attention will be focused on the theoretical properties of the model and on the characteristics of the optimal decision path for land use development.

V. <u>Properties of Optimal Land Use Decisions</u>

The model developed in section 4 can be conceived of as a generalization of a model for environmental preservation developed by Fisher, Krutilla and Cicchetti (1972a, 1972b). A first version of this model for optimal investment decisions was originally developed by Arrow (1968a) and Arrow and Kurz (1970). Extensions introduced into our approach are <u>inter alia</u> : the 0-1 decisions inherent to project evaluations and the possibility of multiple land uses.

The foregoing abstract model for evaluating land use projects will now be employed to derive some theoretical results for optimal decisionmaking as to land use modification. For this purpose only the <u>continuous</u> version of (IV.9) will be further analyzed, while it will be assumed that the additivity condition for land use modification (see also (IV.5)) has been substituted into the relevant functions.

For the sake of clarity the control-theoretic model, adapted to the aforementioned assumptions, is written now as the following continuous Pontryagin optimal control problem (cf. Pontryagin et al.,1962):[2)]

$$\max \omega = \int_0^T \{B(\underline{r},t)-I(t)\} e^{-\rho t} dt$$

s.t.

$$\underline{\dot{r}} = \underline{\varphi} I \qquad (V.1)$$

$$I \geqslant 0$$

Following Arrow (1968a) we postulate positive but diminishing net social benefits on land use developments (in other words, the objective function satisfies the second-order conditions for a maximum).

The Hamiltonian H, associated with the foregoing Pontryagin Model, is equal to :

$$H = e^{-\rho t} \{B(\underline{r},t) - I\} + \underline{\lambda}' \underline{\varphi} I, \qquad (V.2)$$

where $\underline{\lambda}$ is a vector of shadow prices related to the successive land use modifications at time t. The Hamiltonian states that each unit of investment at time t has a twofold effect : (a) an increase in the discounted prospective net benefits (the first term), and (b) a generation of a shadow value for land use modification (the second term). Therefore, the Hamiltonian represents the discounted implicit value of social overhead investments at time t. It is evident, that the control variable $I \geqslant 0$ has to be chosen so as to maximize H.

For the sake of simplicity, (V.2) is rewritten as :

$$H = e^{-\rho t} B(\underline{r},t) + \alpha I, \qquad (V.3)$$

where α is defined as :

$$\alpha = -e^{-\rho t} + \underline{\lambda}'\underline{\varphi} \qquad (V.4)$$

The control variable I appears to enter <u>linearly</u> in the objective function, which may give rise to so-called 'bang-bang' switches (comparable to corner solutions of linear programming models (cf. Bryson and Ho, 1969). Therefore, the properties of the optimum should be carefully inspected.

Owing to the linearity of (V.3) the maximum of H with respect to the control variable is not interior, so that the first-order conditions of H with respect to I do not provide the optimal solution (cf. linear programming models).

It is easily seen that a positive value of α will prevent the solution method from attaining a finite maximum : H would be at a maximum, if $I \to \infty$. Infinite investments over a certain interval, however, would lead to a contradiction with optimal policy, since in this case it would have been more profitable to have invested earlier (so that the policy in the past could not have been optimal). Therefore, the conclusion can be drawn that an optimal policy implies :

$$\alpha \leq 0 \qquad (V.5)$$

It is easily seen that a positive value of I will only prevent a decrease of H, if $\alpha = 0$. On the contrary, if $\alpha < 0$, then I is evidently equal to zero. Therefore, the following conditions should be satisfied for optimal investments for land use modifications :

$$\begin{aligned} \alpha I &= 0 \\ I &\geq 0 \\ \alpha &\leq 0 \end{aligned} \qquad (V.6)$$

The last results can be interpreted as follows. If $\alpha < 0$, then according to (V.4) the aggregated shadow price $\underline{\lambda}'\underline{\varphi}$ of land use investments (i.e., the expected benefit per unit of investment) is lower than the discounted market price of actual capital investments, so that it is not profitable to invest (i.e., I = 0). On the other hand, an indifference as to investing or not investing occurs if $\alpha = 0$. Therefore, the optimal investment path is a 'bang-bang' trajectory, viz. a sequence of intervals satisfying alternately conditions (V.6).

The maximum principle teaches us that the shadow prices $\underline{\lambda}$ are subject to the following evolution path in time :

$$\dot{\lambda} = - \frac{\partial H}{\partial \underline{r}}$$

$$= -e^{-\rho t} \frac{\partial B(\underline{r},t)}{\underline{r}} , \qquad (V.7)$$

with indicates that for each type of land use the changes in the shadow prices of corresponding land use investments should be precisely compensated by the discounted net benefits of the land use modification caused by the investment in question.

Next, (V.4) can be written in differential equation form as :

$$\dot{\alpha} = \rho e^{-\rho t} + \dot{\underline{\lambda}}' \underline{\varphi} \qquad (V.8)$$

$$= e^{-\rho t} \{\rho - \underline{\varphi}' \frac{\partial B(\underline{r},t)}{\partial \underline{r}}\} ,$$

where use is made of (V.7). From a formal point of view the optimal sequence of investment decisions is characterized by (IV.1), (IV.6), (V.6) and (V.8).

In the framework of optimal capital policy the successive intervals of optimal investment decisions represented by (V.6) can be described by means of free and blocked intervals (cf. Arrow, 1968a and Arrow and Kurz, 1970). A blocked interval is characterized by $\alpha < 0$ (i.e. an interval where the non-negativity condition on the investments is active), whereas a free interval is characterized $\alpha = 0$ (i.e., non-negative investments). If the latter condition is valid throughout an interval, it is evident that in a free interval $\dot{\alpha} = 0$, so that according to (V.8):

$$\rho = \underline{\varphi}' \frac{\partial B(\underline{r},t)}{\partial \underline{r}} , \qquad (V.9)$$

which indicates that the <u>aggregated marginal net benefits of modifying</u> a land use pattern with multiple uses is in the optimum equal to <u>the rate of interest</u>. On the other hand during a blocked interval there are no investments, so that the land use pattern remains equal.

The foregoing abstract analysis served to describe the theoretical properties of land use modifications. The assumption was made that an alternative spatial lay-out of a region could be attained by a continuous process of successive development stages. Many infrastructural projects, however, are frequently implemented within a rather limited period of time (for example, when urgent reasons necessitate a rapid implementation of the project concerned). This latter case is essentially characterized by a free interval in which condition (V.9) is valid.

The guidelines, which the optimal control theory may provide to policy-makers, attempt to allocate optimally the available investment funds over the different time periods in order to realise a certain plan, e.g. constructing a canal annex dike. The Dollard-plans, however, do not allow to operationalize fully the theoretical conditions of an optimal control model, because the benefits of any alternative are only measurable after completion of the whole plan. This would imply that during most of the construction time the social net benefits would be zero, and investing in the alternative concerned would be useless, not to say wasting.

Due to the gap between theory and practice we had to restrict our study to a non-dynamic comparison between the performances of the several plans. In the following two sections a more operational analysis of a land use development project in the Dollard Estuary in the northern part of the Netherlands will be described, while an empirical application will be presented in VIII.

VI. Description of the Dollard Projects

The question of the Dollard-canal on which we fill focus the application of cost-benefit and optimal control methods (of section IV and V) reveals several interesting features of changing trends in environmental management.

The Dollard-canal is the crowning piece of a big investment project aimed at improving the drainage system of a substantial agricultural area (90.000 ha = 225.000 acres) in the Northern part of the Netherlands. The main benefit will be a higher agricultural output, of which a high percentage will be processed in the nearby farina-factories and a smaller proportion in the regional straw-board-factories.

On the other hand, implementation of the Dollard-canal according to the original plan would imply a considerable natural sacrifice, because then several dozens of bird species would lose their forage and nesting area.

Only recently, people outside the region became aware of the attack on the environmental area; so that a sharp controversy arose between the administration and the "wildlife-watchers" about the question as to how flexible the government's physical planning policy should be. Is it fair to require that decisions once taken are still subject to debate - even if execution has been started - because the society's value judgements and norms have changed fundamentally ?

This discussion has led the Minister, responsible for Public Works, to revoke the decision taken by one of his predecessors in 1966, inspite of the opposition of many regional authorities. The environmental damage of the original plan was considered too high. Since in this decision process the pros and cons were not explicitly weighted, we thought it worthwile to attempt to measure the implicit evaluation of saving the Dollard-estuary by calculating the additional costs of the alternative chosen compared with the rejected ones.

In 1966 it was decided that through the brackish tidal estuary of the Dollard a so-called offshore drainage-canal annex sea-dike, with a length of about 14 km., had to be constructed, which would embank at the same time 600 ha. of wetlands. The old discharge point in the South-eastern corner of the Dollard (at Nieuwe Statenzijl : point A on map 1) did not function anymore as a consequence of the accretion of land, so that a new discharge point had to be built near deep water (on Cape Reide : B) to use the natural draining potential.

An alternative plan to construct a powerful electric driven pumping-engine at the old discharge point was refused at that time, because then 600 ha. of fertile land would not be reclaimed and the navigation potential of the drainage system would be blocked. Local experts, in addition, opposed the pumping-plan as they needed a guaranteed "all weather" discharge. In fact, the natural-scientific interest was beaten by the economic interest, in spite of the fact that the adopted variant required a higher budget.

To restrict the negative effects on the avifauna the government promised to make a strip uplands along the outside of the new dike that could be used as a new resting, breeding and overtiding area for the birds which are calling at the Dollard.

MAP 1

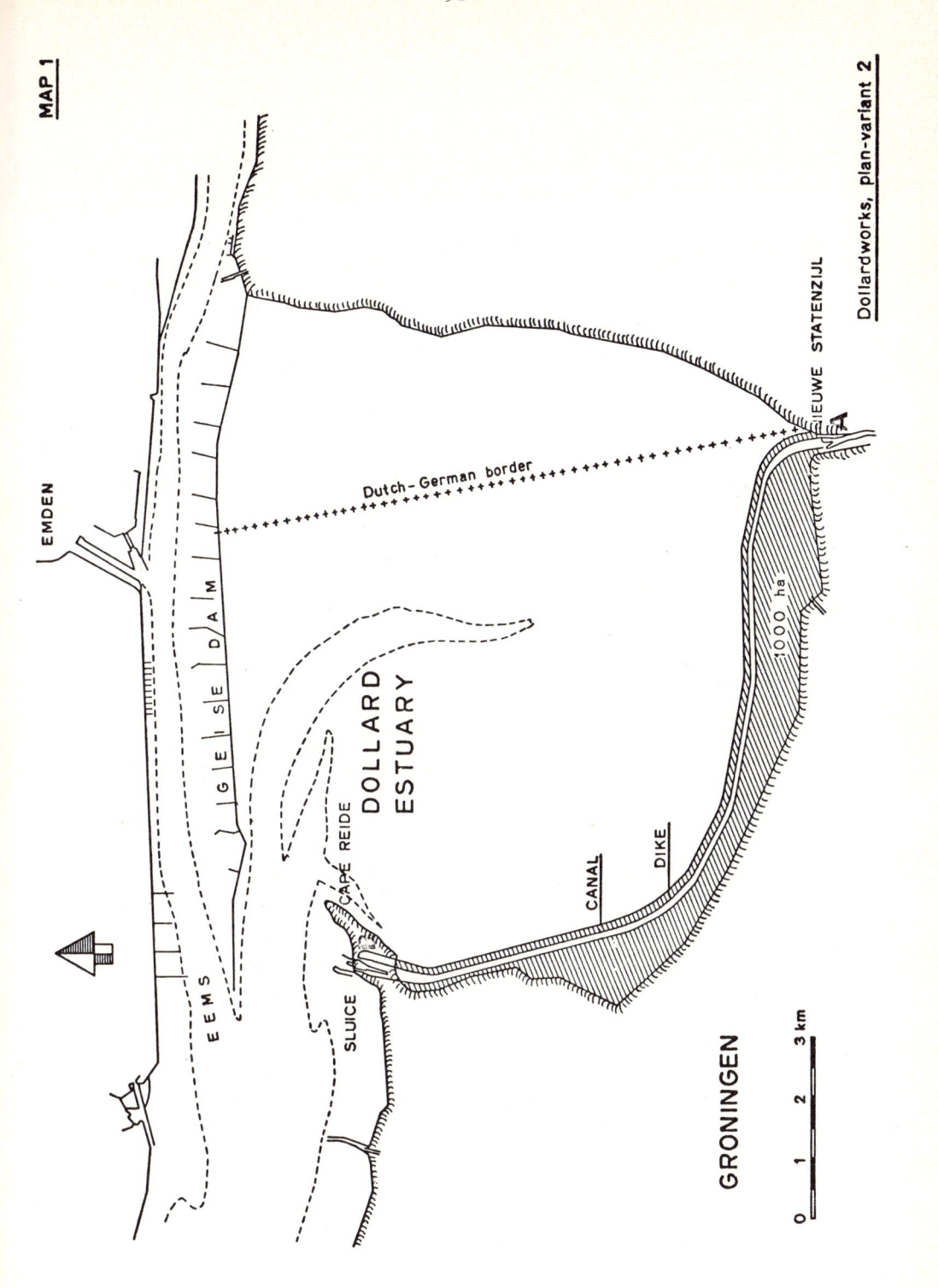

Dollardworks, plan-variant 2

A few years ago the Dollard-works were started by constructing a sluice at Cape Reide; in the meantime this part, in which 20 million guilders are invested, is nearly completed.

The environmental protecting groups have taken the offensive in 1972 in order to prevent the continuation of this canal-plan. They have launched a new alternative consisting of a drainage-canal that is situated not offshore but just at the inside of the old sea-dike. The important natural areas are in this plan unaffected, while free discharge in deep water can occur. The wetlands have not only a feeding function for birds but possess also a very valuable vegetation owing to the brakkish tidal area. This third possibility will use 385 ha. agricultural land, but will save the 400 ha. wetlands earmarked for the canal annex dike and the 600 ha. of the new polder, which would be lost as natural area in the initially accepted plan.

Many provincial authorities have serious objections against the new plan, because they consider the sacrifice of agricultural output too high compared with the advantages of the preservation of the natural environment, and because they do not want to tolerate a lower protection against floods during the construction period. The inside-canal has to cut the old seadike at least even times to evade the irregular form of the sea-wall. But the new dike replacing these sections needs 8 to 10 years before it is stabilized and gives as much protection as the old one did.

The provincial government has reacted to the new proposal of an inside canal with a variant of the original offshore canal by moving the trajectory towards the outside of the old dike. In fact, they developed several variants the trajectory of which was moved on a smaller or greater distance to the coast; we have omitted these variants for the sake of surveyability. Since the most important feeding areas for geese and ducks as well as many high tiding refuges of stilt-walkers are not preserved by these variants, they do not provide a very meaningful contribution to solving the environmental problem.

Briefly summarized, the different alternatives are characterized by the following advantages and disadvantages with respect to the present situation.

Plan 1. **Inside canal**

Advantages : - Improved drainage of 90.000 ha. agricultural land;
- Navigation by barges possible on the drainage canal;
- Complete preservation of all the wetlands in the Dollard Estuary.

Disadvantages:-Loss of 385 ha. first quality arable land, implying a decrease of direct employment for 20 men and lack of scale economies;
- Decreased safety and protection against water floods due to replacing several sections of the existing dike to avoid an irregular route of the canal;
- Postponed improvement in drainage due to the long construction period of 8 to 10 years.

Plan 2. **Offshore canal**

Advantages : - Improved drainage of 90.000 ha. agricultural land;
- Land reclamation of 600 ha. for either agricultural, recreational or natural scientific purposes;
- Navigation by barges possible on the drainage canal;
- Better protection against floods, because a new dike is constructed 600 m. before the coast, so that the old one serves as back-dike.

Disadvantages:-Loss of 1.000 ha. wetlands, so that some birdspecies are severely threatened (of some species 10 to 20 % of the world population stay for some time in the Dollard);
- A rise of the floodmark owing to the decreased basin storage potential of the remaining portion of the Dollard, which may affect the flora and fauna in the creek system.

Plan 3. **Pumping-engine, without drainage canal**

Advantages : - Improved drainage of 90.000 ha. agricultural land, (albeit with more risk than in plan 1 or 2);
- Complete preservation of all the wetlands in the Dollard Estuary;

Disadvantages:-A necessary strengthening and raising of the sea-dike to meet the present standards, which will require an inside strip of about 100 ha. agricultural land.

As mentioned before, the central government has decided that the inside drainage canal will be carried out, putting aside the wishes of the regional and local authorities. This decision is of historical importance, since it will terminate a process of land reclamation in the Dollard which started already in the Middle Ages. In former centuries the generations of that time have wrested from the sea the fertile areas they needed, because nature was subordinated to the production of food. It seems that nowadays the insight begins to break through that maintaining the ecological status quo may be of more importance than sacrificing natural areas for more economic (agricultural) production.

VII. Implicit Evaluation of Environmental Areas by Means of Cost-Benefit Analysis

In the previous section the policy pursued with respect to the development of the Dollard Estuary has been described. In fact, four different development programs could be distinguished.

(a) a policy of maintaining the area in its original state (project 0);
(b) a policy of improving the drainage of the hinterland by digging a canal at the inland side of the sea-dike (project 1);
(c) a policy of improving the drainage of the hinterland by digging a canal at the outside of the sea-bank (project 2);
(d) a policy of improving (partially) the drainage of the hinterland by constructing a pumping-engine with a sufficiently high capacity (project 3).

From these policies only project 2 affects the important ecological area in the Dollard Estuary, viz. the wetlands which serve _inter alia_ as a forage area for birds. The other three projects leave this area almost entirely intact, so that the environmental repercussions of these three projects can be considered to be equal (although the monetary repercussions of these projects are entirely different). A classification of the projects according to their (monetary and environmental) effects is contained in Table 1.

Table 1 : A classification of project effects

	quantifiable monetary effects	environmental effects
project 0	no	no
project 1	yes	no
project 2	yes	yes
project 3	yes	no

It was set out in the previous section that originally a decision had been taken in favour of project 2. Since at that time project 1 was not considered, only a choice had to be made among project 2 and 3 (supposed project 0 was not considered as a relevant one).

Recently, however, project 1 was proposed, and even ultimately selected as the infrastructural plan to be implemented. Now the question we are interested in is : what is the implicit evaluation of the environmental area in question, given the difference in monetary value of the other options in relation to the alternative adopted ? Or more precisely : what are implicitly the net social benefits of maintaining the ecological quality of the area, given the decision to replace the original project 2 by the new project 1 ?

The monetary effects of the projects (mainly consisting of a rise in agricultural productivity due to the improved drainage, the possible loss of agricultural areas, and the regional multiplier effects due to the infrastructural project in question) will be denoted by an upper index M, whereas the intangible environmental effects (loss of coastal wetlands) will be denoted by an upper index E.

The additional net benefits of implementing a certain project n (n=1,2,3) compared with leaving the area in its original state (i.e. plan 0) can be specified now as :

$$\omega_n = \omega_n^M + \omega_n^E \qquad (VII.1)$$

$$= \int_0^T \{B_n^M(t) - C_n^M(t)\} e^{-\rho t} dt + \int_0^T \{B_n^E(t) - C_n^E(t)\} e^{-\rho t} dt$$

Assuming a rational public investment behaviour it is evident that a necessary condition for a decision in favour of a certain project n is :

$$\omega_n = \omega_n^M + \omega_n^E \geq 0, \qquad \forall n \qquad (VII.2)$$

or

$$-\omega_n^E \leq \omega_n^M, \qquad \forall n \qquad (VII.3)$$

Since, however, projects 1 and 3 (the inside canal trajectory and the pumping-engine plan respectively) leave the environmental area uneffected, it is evident that compared to project 0 their net environmental benefits are equal to zero, i.e.,

$$\omega_n^E = 0 \quad , \quad n=1,3 \qquad (VII.4)$$

Furthermore, since project 2 implies a deterioration of an environmental area, one may postulate that the associated net benefits are negative, i.e. :

$$\omega_2^E < 0, \qquad (VII.5)$$

so that ω_2^E represents essentially the net social <u>costs</u> of giving up the environment area.

A second condition for rational behaviour is that the total net social benefits of a selected project should exceed the net social benefits of all other projects. This rule will now be elaborated for the successive decisions taken with respect to the canal project in the Dollard.

Originally, project 2 (the outside canal trajectory, having environmental repercussions) was preferred to project 3 (the pumping engine plan, without environmental repercussions). This implies formally the following implicit condition :

$$\omega_2^M + \omega_2^E \geqslant \omega_3^M + \omega_3^E \quad , \qquad (VII.6)$$

or by taking account of (VII.4) and (VII.5) :

$$-\omega_2^E \leqslant \omega_2^M - \omega_3^M \qquad (VII.7)$$

Since ω_2^E represents the social <u>costs</u> of deteriorating the wetlands it is evident that $-\omega_2^E$ represents the social benefits of maintaining these mud-flats. Condition (VII.7) states that during the first stage of the decision process for constructing a new canal in the Dollard Estuary the social evaluation for the environmental area appeared to fall between 0 and the differential monetary benefits of project 2 with respect to project 3 (assuming that $\omega_2^M > \omega_3^M$). This condition reflects the prevailing ideas at the time that environmental priorities ranked rather low in the social decision process (for a social welfare analysis of such a phenomenon, see Nijkamp and Paelinck, 1973). Since ω_2^M and ω_3^M can in principle be calculated (see VIII), the range of the implicit social evaluation of the environmental area involved can be determined.

More recently, the government took the decision to reject project 2 (in spite of the preparatory operations already implemented) and to accept the newly developed plan of digging a canal at the inland side (i.e. project 1). This implies formally :

$$\omega_1^M + \omega_1^E \geqslant \omega_2^M + \omega_2^E \quad , \tag{VII.8}$$

or taking account of (VII.4) and (VII.5) :

$$-\omega_2^E \geqslant \omega_2^M - \omega_1^M \tag{VII.9}$$

The latter condition states that the social benefits of preserving the environmental area exceed the differential benefits of project 2 with respect to project 1 (assuming that $\omega_2^M > \omega_1^M$). Furthermore, since the public decision-maker took the decision in favour of project 1 owing to its ecological advantages, the total (monetary and environmental) benefits are implicitly deemed to be positive according to condition (VII.2) :

$$\omega_1^M - \omega_2^E \geqslant 0 \quad , \tag{VII.10}$$

where $-\omega_2^E$ represents the (positive) environmental benefits of project 1, viz. a preservation of the coastal wetlands. Condition (VII.10) can also be written as :

$$-\omega_2^E \geqslant -\omega_1^M \tag{VII.11}$$

A comparison of the first range (VII.9) with (VII.11) shows that there will be an overlap of these ranges. In other words, the implicit social evaluation of the coastal wetlands is at least equal to the maximum of the limits represented by (VII.9) and (VII.11). Should this maximum exceed the limit formed by condition (VII.7), then the implicit social evaluation of preserving the environmental area has undergone definitely an absolute rise from the first decision to the other one.

If there is an area of overlap between the last limits, then the previous definite conclusion cannot be drawn (although even in this case a rise in environmental priorities is obvious due to the relatively small area of overlap).

It should be noted that the <u>ex post</u> analysis described above might lead to a negative value of ω_n^M from (VII.2).

The foregoing analysis can be represented as follows :

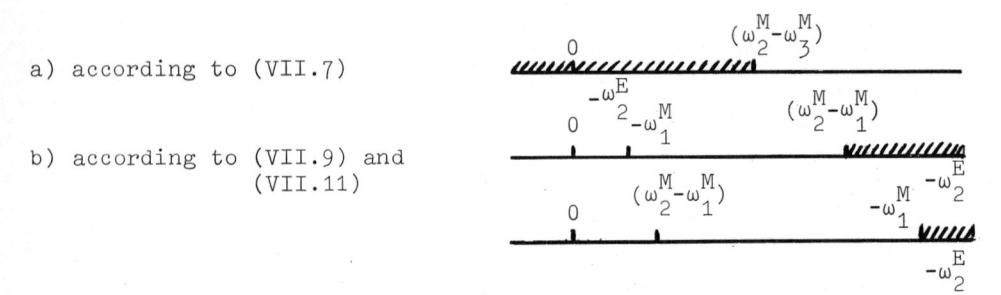

a) according to (VII.7)

b) according to (VII.9) and (VII.11)

Figure 1 : Schematic representation of feasible decision areas

The foregoing analysis will be empirically applied to the Dollard Estuary in the following section.

VIII. Impacts and Monetary Evaluations of the Dollard Plans[3]

In this section an empirical application of the analysis set out in the last paragraph will be presented. First, the quantifiable monetary effects of the plans 1,2 and 3 will be calculated. The monetary impacts of the transformation of land in a canal annex dike, as is the case with the Dollard-works, can be divided into three types of impacts (of section 2), viz. hydrological, physical planning and economic.[4]

The <u>hydrological</u> impacts (H) can be subdivided into the following monetary effects :

H1. Costs of technical constructions;
H2. Costs of special provisions for navigation;
H3. Costs of environmental protection measures;
H4. Benefits of improved drainage for the farms settled in the hinterland areas.

These impacts are summarized in table 2, which is followed by a brief explanation.

Table 2 : Hydrological impacts of the plan-variants of the Dollard-canal annex dike, in millions Dfl

	Plan 1	Plan 2	Plan 3
H1.	-39.3	-34.8	-45.9
H2.	(-8.0)	(-8.0)	-
H3.	- 6.4	- 0.6	-
H4.	+11.5	+16.6	+29.3
Balance	-34.2	-18.8	-20.2

ad H1. The construction outlays are evenly spread over the construction period and discounted back to 1974 at a rate of discount of 9 %. The plan with a relatively long gestation period favours mostly of transforming the original outlays into costs.

Ad H2. The potential use of the canal by barges is not realistic any more at the moment, because the economic size of barges now (2000 tons) is larger than the original depth of the canal (1350 tons), which was optimal at the time of designing, would permit. This is the main reason why the crucial decision of converting the sluice into a lock is postponed continually; in table 2 the amounts of investment costs are only mentioned, but not included in the calculations.

Ad H3. The outlays of environmental protection measures are discounted like H1.

Ad H4. The direct benefits of the improved drainage consist of the present value of the agricultural damage which will be avoided after completion of the project up to 1990 at a rate of discount of 9 %. The damage was in fact a loss of crops on 90.000 hectares agricultural area due to postponed sowing and harvesting because of a too high level of soil water. After completion of the project this damage, which amounted to approximately 2 % of the normal yield of a hectare (see Houdringe Ltd., 1975), will no longer occur.

The indirect and induced effects of these benefits are treated in the paragraph of the economic impacts.

The financial implications with respect to <u>physical planning</u> issues are :

P1. Costs of expropriation of land employed for private use;
P2. Change in value of marginal land;
P3. Infrastructural benefits of the drainage canal for industrial locations in Eastern-Groningen.

The amounts of the different alternatives are presented in table 3.

Table 3 : <u>Physical planning impacts of the plan-variants of the Dollard-canal annex dike, in millions Dfl</u>.

	Plan 1	Plan 2	Plan 3
P1.	-14.6	- 3.0	- 2.5
P2.	-	+ 2.1	-
P3.	x	x	x
Balance	-14.6	- 0.9	- 2.5

Ad P1. The expropriation costs refer to the sales price of agricultural land of Dfl. 25.000 per ha. (in plan 1:585 ha., and plan 3:100 ha.), or of coastal wetlands of Dfl. 7.500 (in plan 2:400 ha.).

Ad P2. The reclaimed land rises in value by Dfl. 6000 per ha. The present value of 600 ha. is Dfl. 2.100.000.

Ad P3. The infrastructural benefits (denoted by 'x') of the drainage canal are becoming less important, because most of continental shipments are performed by trucks. Besides, there are now excellent water transportation facilities available in the newly constructed part in Delfzijl (Eemshaven) at about 50 km. distance (c.f. Werkgroep Dollard, 1973).

The <u>economic</u> impacts include:

E1. The additional, indirect and induced effects in output of the regional industries as a consequence of the original investment-outlays (H1 + H3) for the alternative plan concerned, according to the regional input-output table (cf. Nijkamp and Paelinck, 1975 and Richardson, 1972).
E2. Direct changes in output due to the higher income of the regional

agricultural sector;

E3. Indirect and induced changes in the production-value of other regional industries owing to the increase of regional agricultural output.

Table 4 shows the values of the economic impacts for each plan-alternative.

Table 4 : Economic impacts of the plan-variants of the Dollard-canal annex dike, in millions Dfl.

	Plan 1	Plan 2	Plan 3
E1.	36.1	33.4	33.3
E2.	x	x	x
E3.	0.2	0.3	0.2
Balance	36.3	33.7	33.5

Ad E1. The multiplier effect of the investments is actually the sum of a converging series (extending over a long series of periods) of increased outputs per industry. Here, only the additional output is recorded, since the initial production impulse of the investment-outlay is already accounted for in Table 2. The regional input-output table of 1970 is made available by Federatie van Noordelijke Economische Instituten (1975).

Ad E2. The net change in agricultural output, owing to H4, P1 and P2, causes changes in wages and other income too. No induced effects are calculated, however, because
a) agricultural wages are a substitute for the wages of jobs formerly hold in other industries and
b) the regional distribution of farmers' savings is unknown.

Ad E3. The net increase in agricultural output is a stimulus to increase the output in those industries which are highly dependent upon agricultural inputs. In the region studied, this is only the sector Other Food Processing Industries. According to the assumption of linear and constant technological coefficients of the input-output model, the additional increase in the production-level of this sector in consequence of a rise in agricultural production can be calculated. In the purchasing sector, the earned incomes will also increase, which will imply extra consumption expenditures. If it is assumed that all these extra expenditures are spent in the region itself, one could deduct the multiplier effects of the increased agricultural output. The calculations of these effects are found in Table 5. These effects appear to be rather small.

All monetary impacts are summarized in Table 5.

Table 5 : Total monetary impacts of the plan-variants of the Dollard-canal annex dike, in millions Dfl.

	Plan 1	Plan 2	Plan 3
Hydrological impacts	-34.2	-18.8	-20.2
Physical planning impacts	-14.6	- 0.9	- 2.5
Economic impacts	+36.3	+33.7	+33.5
Total Net Pay-off	-12.5	+14.0	+10.8

The foregoing results will now be analyzed in the framework of section VII. Substituting the net pay-offs of the last two columns in (VII.7) provides the following result :

$$-\omega_2^E \leq 14.0 - 10.8 = 3.2 \qquad (VIII.1)$$

This condition states that at the time that plan 2 was preferred to plan 3, the social costs of deteriorating the wetlands were at most Dfl. 3.2 millions.

The main conclusion of table 5 is deducted from (VIII.9), which gives after substitution :

$$-\omega_2^E \geq 14.0 + 12.5 = 26.5 \qquad (VIII.2)$$

The costs of sacrificing the natural area when plan-alternative 2 would be implemented are now at least 26.5 million guilders, which means that the implicit value the policy-maker assigned to the wetlands is at least Dfl. 26.5 millions. This is the monetary value which made it worthwhile to cancel plan 2, that would destroy the brackish tidal ecosystem, in favour of alternative 1, which preserved this valuable ecosystem. In addition, it may be concluded that the public evaluation of the natural area of the Dollard Estuary has risen considerably between

1966 and 1974. In order to deliminate more accurately the lower limit of the preference area, condition (VII.11) should also be taken into account. Substitution of the results of Table 5 into (VII.11) gives the following result:

$$-\omega_2^E \geq 12.5 \qquad (VIII.3)$$

The latter efficiency condition implies that the evaluation of the wetlands is at least Dfl. 12.5 millions. By combining the range of the value of the natural area from (VIII.2) with the range from (VIII.3) it is easily seen that the implicit public evaluation of preserving the coastal wetlands is at least equal to Dfl. 26.5 millions. This is obviously a significant rise in social value of the environmental area in question.

Only if the benefits of improved drainage (H4.) are calculated for a longer planning period and/or discounted at a lower rate of interest then the net monetary loss of variant 1 may result in a monetary profit. This points to the fact that, besides the valuation of intangible effects, the length of the planning period and the rate of discount have to be determined by the policymaker as well.

To show the interdependency of the magnitudes of these three variables see Table 6, where both the benefits of improved drainage (H4.), and the Total Net Pay-offs (TNP) of alternative 1 are represented, given the values of T and ρ.

Table 6 : Benefits of improved drainage and Total Net Pay-offs of plan-alternative 1, for different values of the planning-horizon and the social rate of discount, in millions Dfl.

Social rate of Discount / Planning-horizon	1990		2000	
	H4	TNP	H4	TNP
$\rho = 0,06$	16.7	-13.6	36.6	6.3
$\rho = 0,09$	11.5	-12.5	22.6	-1.4
$\rho = 0,12$	8.1	-10.6	14.3	-4.4

Table 6 illustrates that the Total (Monetary) Net Pay-off of plan-alternative 1 is positive only if the planning horizon is extended to

2000 and the social rate of discount is 6 %.

A corresponding calculation of the Total Net-Pay off of plan 2 can now be used to determine the lower limits of the environmental evaluations, given the decision in favour of plan 1.

A corresponding calculation of the Total Net Pay-off for plan 2 resulted in an amount of Dfl. 40.6 millions. According to (VII.9) the range of values of the natural area is :

$$-\omega_2^E \geqslant 40.6 - 6.3 = 34.3 \qquad (VIII.4)$$

So, if the Minister of Public Works actually considered a planning-period up to the year 2000 and operated with a rate of discount of 6 %, then his implicit evaluation of preserving the 1000 ha. of coastal wetlands is at least equal to Dfl. 34.300.300, given the decision in favour of the first project.

IX. Conclusions and Perspectives

In the previous sections an attempt has been made to link cost-benefit analysis to optimal control theory. The theoretical analysis showed how the maximum principle could be used to derive optimal investment strategies for land use modifications. The subsequent empirical analysis (partially based on the maximum principle), applied to the Dollard Estuary, appeared to be a useful tool to derive the implicit evaluation of environmental projects, given the decision in favour of one of the alternative plans.

It is evident that it is now worthwhile to extend the scope of this paper by considering the possibility of replacing the ex post analysis by an a priori analysis, so that public decision-making can take into account a whole set of relevant factors related to the project in question. Obviously, the cost-benefit optimal control model can only take account of monetary effects of a certain plan, whereas in reality public investment planning has to consider also the intangibles associated with the project. This implies formally that public investment planning is essentially an evaluation method based on multiple criteria. In other words, the land use problem set out in the previous paragraphs could be specified more adequately as a multi-criterion optimal control problem. An introduction into multi-criterion optimal control problems can be found among others in DaCunha and Polak (1967), Ho (1970), Rao and Rajamani (1975), and Waltz (1967).

Multi-criterion optimal control theory attempts to derive an optimal development path for a dynamic system on the basis of a set of different relevant performance indices. This approach may be illustrated for the continuous version of the optimal control model specified in (V.1) as follows. Suppose that public investments in land use projects are not only guided by the criterion of a maximal net social surplus accruing from the successive investments, but also by the criterion of maintaining or extending the ecological variety of animals, birds and plants within the area concerned. Ecological variety is a variable which reflects the relative frequency of the various types of animals, birds and plants (cf. among others Helliwell, 1969, Hooper, 1971, Van der Maare, 1971, and McHarg, 1969).

Therefore, in addition to the criterion of a maximal net social surplus another performance index, viz. a maximal ecological variety (ν), can be introduced, i.e.,

$$\max \int_0^T \nu \, dt \qquad (IX.1)$$

It is evident, that the ecological variety is mainly determined by the various types of land use and the amount of public investment. Therfore, ν can be formalized in differential equation form as a function of $\underline{r}$ and I :

$$\dot{\nu} = f(\underline{r}, I) \qquad (IX.2)$$

Finally, the total multi-criterion optimal control model can now be written as :

$$\max \omega_1 = \int_o^T \{B(\underline{r},t) - I(t)\}e^{-\rho t} dt$$

$$\max \omega_2 = \int_o^T \nu \, dt$$

$$\underline{\dot{r}} = \varphi \, I$$

$$\dot{\nu} = f(\underline{r}, I) \qquad\qquad (IX.3)$$

$$I \geqslant 0$$

$$\underline{r} = \underline{r}_o, \; \nu = \nu_o$$

The latter bicriterion control model can be solved in different ways. A first method is to select <u>a priori</u> one <u>dominant</u> criterion functional (either ω_1 or ω_2) and to impose a lower limit on the other one. The shadow price from the maximum principle associated with such a lower limit reflects the relative evaluation of a certain lower limit, so that such a shadow price forms a basis for a trade-off among the criterion functionals.

Another solution method is to seek for a so-called <u>efficient</u> solution, which essentially represents a Pareto-optimum, so that none of the objective criteria can be increased without decreasing one of the others. Obviously, a unique solution is not guaranteed here, but the method gives at best a (limited) set of feasible appropriate solutions among which ultimately a political choice has to be made.

It is evident, that the latter field of multi-criterion optimal control theory is still rather undiscovered. The authors are convinced, however, that these methods possess a large perspective, so that more attention for this field is definitely worthwhile.

Footnotes

1. An introduction into optimal control theory can be found among others in Athans and Falb (1966), Bryson and Ho (1969), Kirk (1970), Lee and Markus (1967), Luenberger (1970) and Tabak and Kuo (1971), whereas economic interpretations and applications are contained among others in Arman (1968), Arrow (1968a, 1968b), Arrow and Kurz (1970), Benavie (1970), Jacquemin and Thisse (1972), and Peterson (1973). Applications of control theory to environmental problems are contained among others inf Fisher, Krutilla and Cicchetti (1972a, 1972b), Herfindahl and Kneese (1974), Nijkamp and Paelinck (1973) and Nijkamp (1974a).

2. The model is assumed to be relevant for each plan n : it describes the optimal time-varying path for each plan n. For the sake of brevity the lower index n will be omitted in the relationships.

3. The authors are really much indebted to Anje Valk, who collected many obscure data and performed the computer-calculations, and to Frans Kutsch Lojenga for his assistance.

4. Detailed calculations of the project are available from the authors on request.

References

Armand, R., 1968, "Interpretation Economique du Principle du Maximum de Pontryagin", Metra, 7, no. 3, 495-548.

Arrow, K.J., 1963, Social Choice and Individual Values, 2nd ed. (Wiley, New York).

Arrow, K.J., 1968a, "Optimal Capital Policy with Irreversible Investment" in, Value, Capital and Growth (J.N. Wolfe e.d.), (Edinburgh University Press, Edinburgh),1-20.

Arrow, K.J., 1968b, "Applications of Control Theory to Economic Growth" in, Mathematics of the Decision Sciences (G.B. Dantzig and A.F. Veinott, eds.) (Providence, Rhode Island),85-119.

Arrow, K.J. and M. Kurz, 1970, "Optimal Growth with Irreversible Investment in a Ramsey Model", Econometrica, 38, no.2, 331-344.

Athans, M. and P.L. Falb, 1966, Optimal Control; An Introduction to the Theory and Its Applications (McGraw-Hill, New York).

Benavie, A., 1970, "The Economics of the Maximum Principle", Western Economic Journal, 7, 426-430.

Bergson, A., 1975, "On the Concept of Social Welfare", Quarterly Journal of Economics, 68, 233-252.

Bryson, A.E., and Yu-Chi Ho, 1969, Applied Optimal Control (Blaisdell Publishing Co., Massachusetts).

DaCunha, N.O., and E. Polak, 1967, "Constrained Minimization under Vector-valued Criteria in Finite Dimensional Spaces", Journal of Mathematical Analysis and Applications, 19, 103-124.

Dupuit, J., 1844, "De la Mesure de l'Utilité des Travaux Publics", Annales des Ponts et Chaussées.
Translated by R.H. Barback, 1952, "On the Measurement of the Utility of Public Works", International Economic Papers, no. 2, 83-110.

Federatie van Noordelijke Economische Instituten, 1975, "De komst van de Centrale Directie der P.T.T. : Enkele economische gevolgen voor het Noorden des Lands", Groningen, Bijlage D.

Fisher, A.C., J.V. Krutilla, and C.J. Cicchetti, 1972a, "The Economics of Environmental Preservation : A Theoretical and Empirical Analysis", American Economic Review, 62, 605-619.

Fisher, A.C., J.V. Krutilla, and C.J. Cicchetti, 1972b, "Alternative Uses of Natural Environments : The Economics of Environmental Modification", in Natural Environments : Studies in Theoretical and Applied Analysis (J.V. Krutilla, ed.), (John Hopkins Press, Baltimore), 18-53.

Friedlaender, A.F., 1973, "Macro Policy Goals in the Postwar Period : a Study in Revealed Preference", Quarterly Journal of Economics, vol. 87, 25-43.

Helliwell, D.R., 1969, "Valuation of Wildlife Resources", Regional Studies, 3, 1969, 41-47.

Herfindahl, O.C. and A.V. Kneese, 1974, Economic Theory of Natural Resources (Merrill Publishing Cy, Columbus).

Ho, Y.C., 1970, "Differential Games, Dynamic Optimization and Generalized Control Theory", Journal of Optimization Theory and Application, 6, no. 3, 179-209.

Hooper, M.D., 1971, "The Size and Surroundings of Nature Reserves", in, The Scientific Management of Animal and Plant Communities for Conservation (E. Duffey, and A.S. Watt, eds.) (Blackwell Scientific Publishing Co., Oxford), 555-561.

Houdringe, 1975, "Raming van schade wegens optredende afvoerstagnaties voor de landbouw", De Bilt, Grontmij N.V.

Jacquemin, A.P. and J. Thisse, 1972, "Strategy of the Firm and Industrial Environment : An Application of Optimal Control Theory", Working Paper no. 7205, University of Louvain.

Kirk, D.E., 1970, Optimal Control Theory; An Introduction, Prentice-Hall, Englewood Cliffs.

Klaassen, L.H., A.C.P. Verster and T.H. Botterweg, 1974, Kosten-baten-analyse in regionaal perspectief (Tjeenk-Willink, Groningen).

Lee, E.B., and L. Markus, 1967, Foundations of Optimal Control Theory, (Wiley, New York).

Little, I.M.D., 1952, "Social Choice and Individual Values", in Journal of Political Economy, 60, no. 5, 422-432.

Luenberger, D.C., 1970, "Mathematical Programming and Control Theory : Trends of Interplay", Paper 7th International Symposium on Mathematical Programming, The Hague.

Maarel, E. van der, 1971, "Florastatistieken als Bijdrage tot de Evaluatie van Natuurgebieden", Gorteria, 5, 176-188.

McHarg, I.L., 1969, Design with Nature, Natural History Press, New York.

Nijkamp, P., 1974a, "Spatial Interdependencies and Environmental Effects", in, Dynamic Allocation of Urban Space (A. Karlqvist, L. Lundqvist and F. Snickars, eds.) (Saxon House, Farnborough),175-209.

Nijkamp, P., 1974b, "A Multi-criteria Analysis for Project Evaluation; Economic-ecological Evaluation of a Land Reclamation Project", Paper presented at the Annual North-American Meeting of the Regional Science Association, Chicago, Research Memorandum No. 10, Department of Economics, Free University, Amsterdam.

Nijkamp, P., 1975, "Operational Determination of Collective Preference Parameters", Research Memorandum no. 17, Department of Economics, Free University, Amsterdam.

Nijkamp, P., and J.H.P. Paelinck, 1973, "Some Models for the Economic Evaluation of the Environment", Regional and Urban Economics, 3, no. 1, 33-62.

Nijkamp, P., and J.H.P. Paelinck, 1975, Operational Theories and Methods in Regional Economics (Saxon House, Farnborough).

Nijkamp, P., and W.H. Somermeyer, 1971, "Explicating Implicit Social Preference Functions", The Economics of Planning, 11, no. 3, 101-119.

Paelinck, J.H.P., 1973, "Collective Preference Functions Revisited", in Series : Foundations of Empirical Economic Research, Netherlands Economic Institute, Rotterdam.

Peterson, D.W., 1973, "The Economic Significance of Auxiliary Functions in Optimal Control", International Economic Review, 14, 1, 1973, 234-252.

Pontryagin, L.S., V.G. Boltyanskii, R.V. Gamkrelidze, and E.F. Mischenko, 1962, The Mathematical Theory of Optimal Processes, (Interscience, New York).

Rao, P.K., and V.S. Rajamani, 1975, "A New Approach to Public Investment", Socio-Economic Planning Sciences, 9, 11-14.

Rausser, G.C., and J.W. Freebairn, 1974, "Estimation of Policy Preference Functions : An Application to U.S. Beaf Import Quotas", Review of Economics and Statistics, 56, no.4, 437-449.

Richardson, H.W., 1972, Input-Output and Regional Economics (Weidenfeld and Nicolson, London).

Simon, H.A., 1958, Models of Man : Social and Rational, Wiley, New York.

Tabak, D., and B.C. Kuo, 1971, Optimal Control by Mathematical Programming (Prentice-Hall, Englewood Cliffs).

Waltz, F.M., 1967, "An Engineering Approach to Hierarchical Optimization Criteria", IEEE Transactions Automatic Control, AC-12, no. 2, 179-180.

Warnke, D.W., N.C. Terre and A.P. Ameiss, 1973, "A Methodology for Determining Public Investment Criteria", Socio-Economic Planning Sciences, 7, 317-326.

Werkgroep Dollard, 1973, "De Dollard bedreigd", Aanvullende Nota, Harlingen.

ENVIRONMENT AND POPULATION OPTIMUM

Bernhard Felderer
University of Köln
Köln - W.Germany

I. Introduction

The paper attempts to reformulate the old problem of the optimum size of population in the context of negative externalities and environmental factors. Historically two approaches to this problem can be distinguished.

The first approach has its roots in the ideas of J.S. Mill and was later developed by Cannan (Cannan, 1888) and K. Wicksell (Wicksell, 1933). The basic model is the so-called classical production function with first increasing and then decreasing returns to labor while the other factors of production are held constant. Already these early authors have defined that the size of the population is optimum when per-capita production (income) is at its maximum. At a first glance, the constancy of capital and natural resources in a long-term problem is such a strong restriction that the possibility of any relevant conclusion seems doubtful. Contrary to this we have to consider that the production function would also be of the classical or neoclassical type if labor and capital would be variable but natural resources and technology constant.

The second approach to this problem is not the maximization of a per-capita-production function but the maximization of a social welfare function (Penrose, 1934). We do not intend to discuss the problem of the logical existence of a social welfare function in the sense of Arrow's "possibility theorem" (Arrow, 1951) or the problem of the determination of such a function in reality. But we have to examine the stability of such a preference ordering in a long-term economic problem. Do human beings have ideas about optimal, geographical and cultural spaces of freedom not variable over time, which necessarily have to be narrowed down with an increasing population ? Is their adaptability of such an extent that especially by the change of generations an adaptation of preferences to very high population densities is possible ? However, the preferences of the individual concerning the optimum density of the population are determined essentially by the population density and the social environment in which he lives. Thus individual preferences will be highly variable - especially through the change of generations - which makes the instability of the social welfare function over time obvious. Therefore we conclude that the use of a social welfare function in considerations about the population optimum is inappropriate.

H.L. Votey (1969) has combined the per-capita production function with a welfare function which implies a trade-off between increasing utility by increasing population and increasing utility by higher per-capita income. Votey's article contains several mathematical errors (Compare : Russel, 1972). To justify the assumption that utility increases with increasing population, Votey argues that it is an obvious fact that having children increases utility. He disregards the fact that the decision - we assume the rationality of this decision - to have children is made at any moment or in any time period by only a small part of the population. While for this part the additional children will bring about a higher utility, they may at the same time represent a decrease of utility for the rest of the population. Thus we cannot accept Votey's welfare function. Actually we here have a typical external effect. To have children and thus to increase the population also affects the utility of all the other members of the economy. By this consideration we have found a suitable transition to the present model.

II. The static population optimum

The basic idea is the following : in a closed economy each additional individual will induce a positive production effect and a negative cost effect. The cost effect results from the negative external effects which are caused by the existence of an additional human being. A closed region or economy offers only a limited capacity of regeneration of polluted "primary" - non reproducible - resources like water, air and recreation areas. Generally it will be safe to assume that the capacity of an urban area to renew these resources is too small. Thus the question as to the size of the required region of regeneration for a given urban area has not yet been answered. What we do know is that the quality of these resources has deteriorated in the last years in many urban areas and in whole industrialized countries.

The model presented below follows the first of the two approaches mentioned in the introduction above. However, it is different in the sense that the population optimum is determined not only by a production function but also by a function of negative external effects.

The production function which first displays increasing and then decreasing returns to the variable factor is - as usual - called the classical production function.

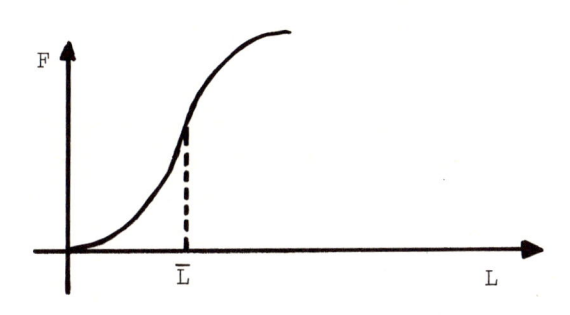

Figure 1

For our purposes it is sufficient to consider from all possible factors of production only capital and labour :

$$F = F(K,L) \qquad (1)$$

where F denotes production;
K capital and L labor.

K and F are expressed in monetary terms, L in labor time. Function (1) is continuous and the first derivatives are differentiable everywhere. Equation (1) has the following properties :

A quantity of the variable factor at which $\frac{\partial F}{\partial L} = 0$ does not exist for finite values of L : $\lim_{L \to \infty} \frac{\partial F}{\partial L} = 0$.

Furthermore :

$$\frac{\partial F}{\partial L} > 0 \qquad (2)$$

$$\frac{\partial^2 F}{\partial L^2} > 0 \quad \text{for} \quad 0 \leqslant L < \bar{L} \tag{3}$$

$$\frac{\partial^2 F}{\partial L^2} < 0 \quad \text{for} \quad L > \bar{L} \tag{4}$$

In addition we require : $F \geqslant 0$, $K \geqslant 0$, $L \geqslant 0$.

These properties hold also for the second factor of production. The independent variables in the function of negative external effects are - as in the production function - K and L. The idea is that an increase of production capacity when technology is constant will deteriorate resources like water, air etc. just as an increase of population.

The non-linearity of the function can be justified as follows. In the case of small population sizes, the only costs are those to avoid the deterioration of "primary" resources, generally by technical appliances at the stage of production. With increasing population size we have to take into account additional costs which are the result of pollution of water and air etc., like medical treatment of diseases caused by pollution, inability to work, costs of early retirement etc. There are costs which exist only above a certain density of population. Then the recreation areas situated around an urban center can loose their function because too many people are using them. The result would be additional transportation and time costs in order to reach recreation areas at greater distance from the urban center.

If the population size reaches the value L^* the negative external effects are G^*. At this point the quality of the non-reproducible "primary" resources like water, air etc. has become so bad that human life in this closed region or economy is no longer possible.

The general civil use of gas-masks in case of air pollution alarm in certain urban areas of the world, gives to the point L^* a practical meaning.

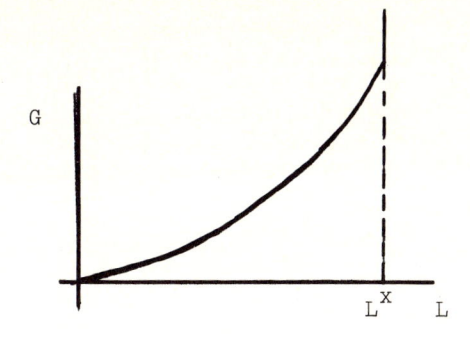

Figure 2

Consequently we define the following relation :

$$G = G(K,L) \tag{5}$$

where G is the value of the negative external effects. A measurement of these effects in monetary terms would be possible for example by estimating the costs for measures necessary to avoid these effects.

The function (5) has the properties :

$$\frac{\partial G}{\partial L} > 0 \quad \text{for } L > 0 \tag{6}$$

$$\frac{\partial^2 G}{\partial L^2} > 0 \tag{7}$$

$$\lim_{L \to L^X} \frac{\partial G}{\partial L} = \infty \quad \text{and } 0 < L^X < \infty \tag{8}$$

And we also require : $G \geqslant 0$, $K \geqslant 0$, $L \geqslant 0$

The subtraction of the function G from F defines our "net production function" Q :

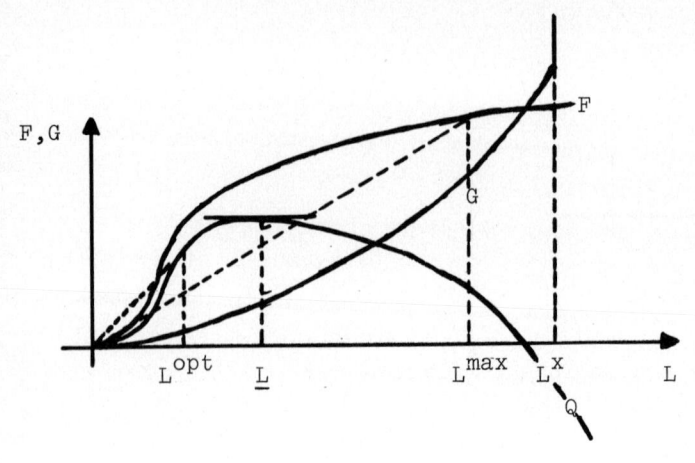

Figure 3

We define :

$$Q = F(K,L) - G(K,L) \qquad (9)$$

or

$$Q = H(K,L) \qquad (9a)$$

As we have argued in the introduction above, we define the optimum size of population as the size at which the per-capita income (= production) is at its maximum. We consider as the production function our net function H and not - as usual - the function F. We assume that the notions labor force and population are identical. This simplification could be eliminated by the introduction of a suitable coefficient which we omit in order to restrict the ideas to their essentials.

$$\frac{Q}{L} = \frac{1}{L} \cdot H(K,L) \qquad (10)$$

The necessary condition for a maximum is that $\frac{\partial (\frac{Q}{L})}{\partial L}$ equals zero. We assume that this condition is also sufficient for a maximum and we get :

$$\frac{\partial H}{\partial L} = \frac{H}{L} \qquad (11)$$

If we maximize output per head $\frac{Q}{L}$:

$$\frac{Q}{L} = \frac{1}{L} \cdot F(K,L) - \frac{1}{L} \cdot G(K,L), \qquad (12)$$

we get the following result :

$$\frac{\partial F}{\partial L} - \frac{\partial G}{\partial L} = \frac{F}{L} - \frac{G}{L} \qquad (13)$$

The results (11) and (13) are formally equivalent. The differences of the marginal and average products of the functions F and G correspond to the marginal and average product of function H. This static optimum can be shown graphically by a tangent from the origin to the net production function.

The population optimum L^{opt} has to be distinguished from the population maximum L^{max}. The maximum is defined by the fact that income or production per-capita cannot fall below a given subsistence level $(\frac{F}{L})_{min}$. Up to the point L^x the function G can be neglected because we consider here only the problem of survival of a maximum number of people. Because of the definition of L^x given above the population size cannot be greater than L^x. The population maximum can be demonstrated graphically as the intersection of a straight line through the origin with the slope $(\frac{F}{L})_{min}$ and the production function F.

Each economy has the alternative to use net investment to enlarge its capital stock (K_p) - that means she will move along a given production function - or to use investment for a reduction of negative external effects (K_u). This is identical with a shift downwards of function G and an increase of restriction L^x. Thus we have the following production function :

$$\frac{Q}{L} = \frac{1}{L} \cdot F(K_p,L) - \frac{1}{L} \cdot G(K_p,L) \cdot f(K_u) \qquad (14)$$

where $f(K_u)$ has the properties :

$$f(0) = 1$$
$$\frac{\partial f}{\partial K_u} < 0$$
$$\frac{\partial^2 f}{\partial K_u^2} > 0 \qquad (15)$$
$$\lim_{K_u \to \infty} f(K_u) = 0$$

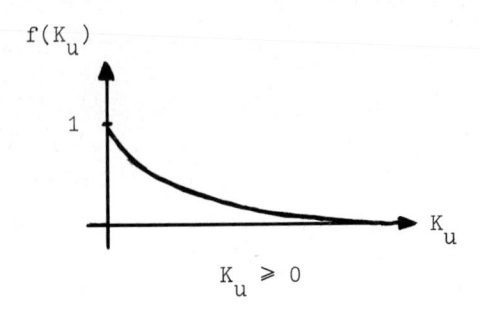

Figure 4

Because the total of the capital in the production process (K_p) and the capital employed to reduce pollution is limited, we can find the population optimum by maximizing function (14) with the subsidiary condition $K_p + K_u = K$. To this end we use the following Lagrangian:

$$\mathcal{L} = \frac{1}{L} \cdot F(K_p, L) - \frac{1}{L} \cdot G(K_p, L) \cdot f(K_u) + \lambda(K_p + K_u - K) \qquad (16)$$

Necessary conditions for a maximum are that the first derivatives of (16) with respect to K_p, K_u, L and λ are equal to zero. Solving these four equations simultaneously we get:

$$\frac{\partial F}{\partial L} - f \cdot \frac{\partial G}{\partial L} = \frac{F}{L} - f \frac{G}{L} \qquad (17)$$

and

$$\frac{\partial F}{\partial K_p} - f \frac{\partial G}{\partial K_p} = - G \cdot \frac{\partial f}{\partial K_u} \qquad (18)$$

The result of equation (17) corresponds to that of equations (11) and (13). When the size of the population is at its optimum the marginal product and the average product of labor of the net production function have to be equal. Equation (18) can be interpreted as follows.

The disposable capital has to be allocated to the two alternatives, production and reduction of negative externatlities, such that the value of the marginal product of capital employed in production is equal to the marginal value product which results from a reduction of negative externalities.

III. The optimal size of population and technical progress

Technical progress is always defined as an upward shift of the production function. Any statement about a change in the population optimum (L^{opt}) is possible only if we know precisely how the function shifted. This shift firstly depends on the definition of the type of technical progress and then on the values of the relevant parameters.

We assume here the existence of so-called Hicks-neutral technical progress :

$$F = F(K,L) \cdot h(t) \qquad (19)$$

where h is the progress function and t the time variable. We choose the following explicit form of h :

$$F = F(K,L) \cdot e^{\lambda t} \qquad (19a)$$

with λ as constant rate of progress and $\lambda > 0$.

In order to find the optimal size of the population we have to maximize the following function according to the definition given above :

$$\frac{Q}{L} = \frac{1}{L} \cdot F(K,L) \cdot e^{\lambda t} - \frac{1}{L} \cdot G(K,L) \qquad (20)$$

The necessary condition for a maximum of (20) with respect to L yields :

$$\frac{\partial F}{\partial L} \cdot e^{\lambda t} - \frac{\partial G}{\partial L} = \frac{F}{L} \cdot e^{\lambda t} - \frac{G}{L} \qquad (21)$$

from which we obtain

$$L = \frac{F \cdot e^{\lambda t} - G}{\frac{\partial F}{\partial L} \cdot e^{\lambda t} - \frac{\partial G}{\partial L}} \qquad (22)$$

We now have to examine the problem of how L^{opt} will change with an increasing t. This question can be solved with a simple limit :

$$\lim_{t \to \infty} L^{opt} = \frac{F}{\frac{\partial F}{\partial L}} \qquad (23)$$

Using empirical values for F, λ, G, $\frac{\partial F}{\partial L}$ and $\frac{\partial G}{\partial L}$ we find that L^{opt} converges within a few periods to the value $F(\frac{\partial F}{\partial L})^{-1}$. Depending on the values of F, G, $\frac{\partial F}{\partial L}$ and $\frac{\partial G}{\partial L}$ the limit $F(\frac{\partial F}{\partial L})^{-1}$ is approximated from below or above.

We conclude from (23) that the optimum size of the population stays contant in spite of Hicks-neutral technical progress.

We now examine the question if under the assumption of Hicks-neutral technical progress the population maximum will also be constant. We have defined above the maximum size of the population (L^{max}) as the size at which per-capita income is equal to $(\frac{F}{L})_{min}$. The function G has an influence on the determination of the maximum only if L is equal to the restriction L^x.

$$\frac{F(K,L)}{L^{max}} \cdot e^{\lambda t} = (\frac{F}{L})_{min} \qquad (24)$$

or

$$L^{max} = \frac{F(K,L)}{(\frac{F}{L})_{min}} \cdot e^{\lambda t}$$

The quantity of factors of production being constant, the population maximum increases through Hicks-neutral technical progress at the rate λ. It follows that, if technical progress in reality can be described as defined above, the difference between population optimum and maximum will constantly increase as a consequence of progress.

The question of how technical progress which reduces negative externalities influences the population optimum has still to be examined. To this end equation (20) is written as follows :

$$\frac{Q}{L} = \frac{1}{L} \cdot F(K,L)e^{\lambda t} - \frac{1}{L} \cdot G(K,L) \frac{1}{e^{\gamma t}} \qquad (25)$$

where γ is a constant rate of reduction of negative external effects. We require : $\gamma > 0$.

From the necessary condition for a maximum of (25) we get :

$$L^{opt} = \frac{Fe^{\lambda t} - G \frac{1}{e^{\gamma t}}}{\frac{\partial F}{\partial L}e^{\lambda t} - \frac{\partial G}{\partial L}\frac{1}{e^{\gamma t}}} \qquad (26)$$

And finally as above in (23) :

$$\lim_{t \to \infty} L^{opt} = \frac{F}{\frac{\partial F}{\partial L}} \qquad (27)$$

Thus a reduction of negative externalities by Hicks-neutral technical progress does not - after convergence - influence the population optimum.

In equation (14) we assumed that an economy has the alternative to invest its capital in production or in the reduction of pollution, etc. Following the Lagrangian of (16) we also would like to examine the question if and how the population optimum is influenced by Hicks-neutral technical progress. Equation (16) has to be written as :

$$\mathcal{L} = \frac{1}{L} F(K_p, L)e^{\lambda t} - \frac{1}{L} G(K_p, L)f(K_u) + \lambda(K_p + K_u - K) \qquad (28)$$

We obtain as solution for the population optimum :

$$L^{opt} = \frac{F e^{\lambda t} - f \cdot G}{\frac{\partial F}{\partial L}e^{\lambda t} - f\frac{\partial G}{\partial L}} \qquad (29)$$

If the factors of production are held constant and t increases L^{opt} converges very fast to a constant value :

$$\lim_{t \to \infty} L^{opt} = \frac{F}{\frac{\partial F}{\partial L}} \qquad (30)$$

Thus again, Hicks-neutral technical progress does not change the population optimum.

For these considerations we have used a very general definition of technical progress. However, progress in reality might be different form our definition. If technical progress is non-neutral in the sense of Hicks, then the changes of the parameters of the function have to be known in order to make statements about the population optimum possible. The optimum could increase as well as decrease.

IV. Population and economic growth

In the last part of the paper we examine the dynamic aspects of the problem.

A population cannot grow indefinitely at a constant rate as it is assumed in the typical neoclassical model. This would mean that the population size also approaches infinity.

$$L(t) = L_o \cdot e^{G_L t}$$

where G_L is the growth rate of labor and t the time variable $L_o > 0$, $G_L \geq 0$.

Now, if $t \to \infty$, $L \to \infty$.

If we exclude that $L \to \infty$ then G_1 can only be zero. If G_L is variable then it will have to go to zero.

A dynamic model should therefore not try to determine a positive, optimal growth rate of the population (Phelps, 1967), but should rather demonstrate the mechanism by which this rate will finally go to zero. We will follow this approach using well known assumptions of neoclassical growth theory.

As the population is treated as an endogenous variable we have to make assumptions about the population growth. The best known assumption is the so-called classical population theory according to which the population growth rate is proportional to the difference between the current per-capita-income and the subsistence level. We adopt here a special version of this theory (Nelson, 1956). We assume that the death rate is dependent on income up to a given income level y^1. The birth rate is independent from income.

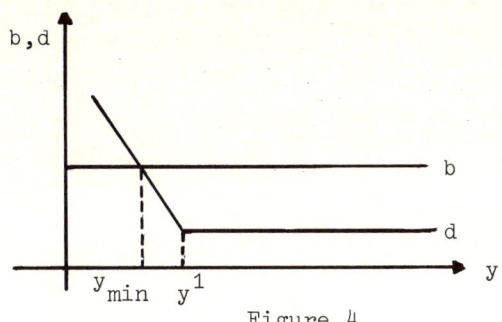

[Figure 4: graph showing b and d vs y, with b constant and d decreasing then constant, crossing at y_{min} and leveling at y^1]

Figure 4

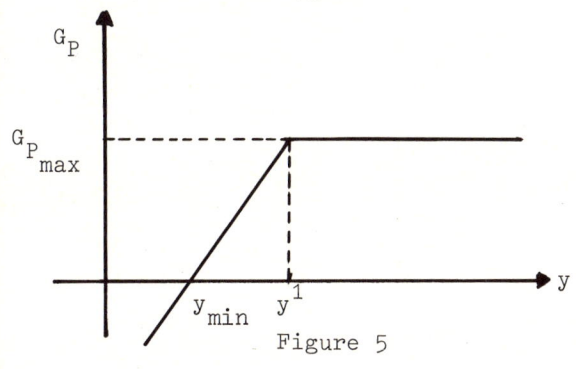

[Figure 5: graph showing G_P vs y, linear from y_{min} up to y^1 where it reaches $G_{P_{max}}$ and stays constant]

Figure 5

where b = birth rate
d = death rate

y_{min} is the per-capita-income at which the population is constant (b=d).

If $y < y_{min}$ ⇒ $G_P = \frac{\dot{P}}{P} < 0$

If $y > y_{min}$ ⇒ $G_P > 0$

This can be expressed by the following equations:

$$G_P = p(y - y_{min}) \quad \text{for } y < y^1$$
$$G_P = G_{P_{max}} \quad \text{for } y \geq y^1 \quad (31)$$

We use a production function of the neoclassical type:

$$Y = a\, K^\alpha\, L^{1-\alpha} \cdot e^{\lambda t} \quad (32)$$

where Y is the production
λ is the rate of technical progress

t is the time variable

This Cobb-Douglas function is linear-homogenous with technical progress which is neutral in the sense of Hicks, Harrod and Solow.

$$L = l\, P \qquad (33)$$

where P = Population

The labor force is always a given constant fraction of the population.

$$G_K = \frac{\dot{K}}{K} = \frac{I}{K} \qquad (34)$$

where G_K = growth rate of capital

This equation defines the growth rate of capital as the relation of net investment to the capital stock.

$$I = S \qquad (35)$$

Equation (35) is the usual equilibrium condition.

$$S = s(y - y_o)P \qquad (36)$$

Saving is a linear function of the difference between a given per-capita-income and the per-capita-income at which saving is zero. For the sake of simplicity we assume that $y_o = y_{min}$.

Firstly we have to examine if a constant equilibrium growth rate exists. From equation (31) follows for the per-capita-income (y) :

$$y = l \cdot a\, (\tfrac{K}{L})^\alpha\, e^{\lambda t} \qquad (37)$$

From this equation and (34), (35), (36) and (33) we get :

$$G_K = s \cdot l^{\frac{1-\alpha}{\alpha}} \cdot (\tfrac{a}{y})^{\frac{1}{\alpha}} \cdot e^{\frac{\lambda}{\alpha}t} \cdot (y - y_{min}) \qquad (38)$$

Graphically, function (38) can be represented as follows :

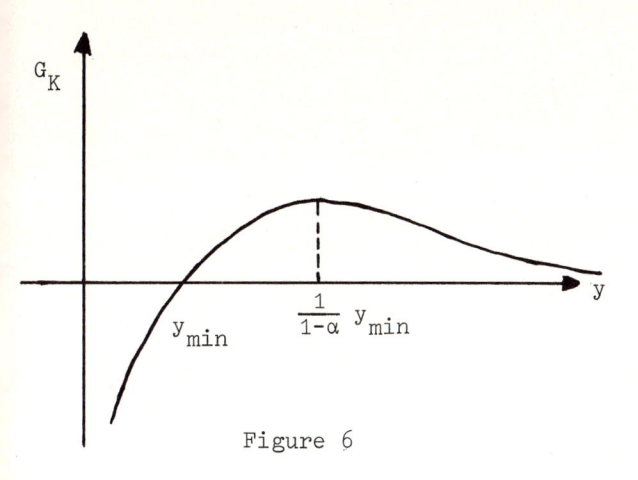

[Figure 6: Graph of G_K vs y, with y_{min} and $\frac{1}{1-\alpha}y_{min}$ marked on the y-axis]

Figure 6

For $y \to 0$ we have from (38) $\Rightarrow G_K \to -\infty$

For $y \to \infty \Rightarrow G_K \to 0$

For $y = \frac{1}{1-\alpha}y_{min}$ the function is at its maximum.

Equation (31) and (38) can now be used to demonstrate graphically the existence of equilibrium solutions :

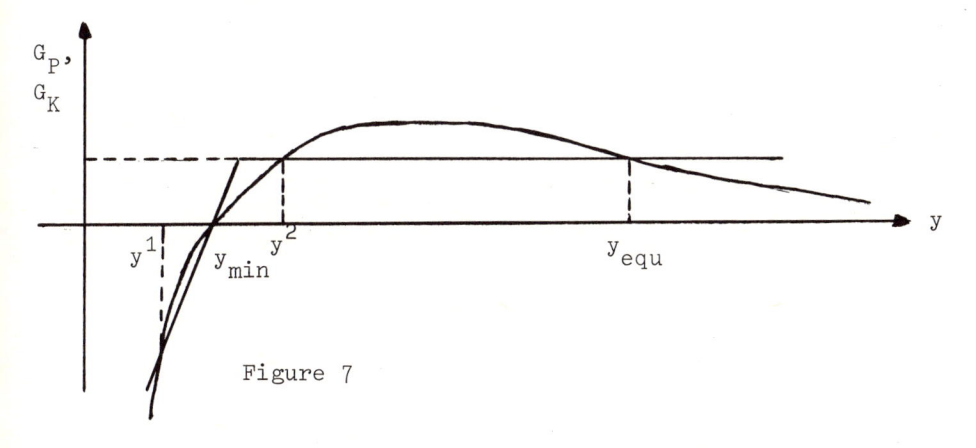

[Figure 7: Graph of G_P, G_K vs y, with y^1, y_{min}, y^2, and y_{equ} marked]

Figure 7

Given our function (31) there can be up to 4 intersections of equations (31) and (38). We have two stable equilibrium solutions : For $y > y_2$ the equilibrium is y_{equ}. For $y_1 < y < y_2$ the equilibrium is y_{min}. The existence of a second - lower - stable equilibrium has been called

"Low-Level Equilibrium Trap" (Nelson, 1956). However, the growth rate of capital (38) is dependent also on the time variable and equation (38) is constantly moved upwards with the rate $\frac{\lambda}{\alpha}$. Thus the low-level equilibrium in y_{min} will exist for a short time only.

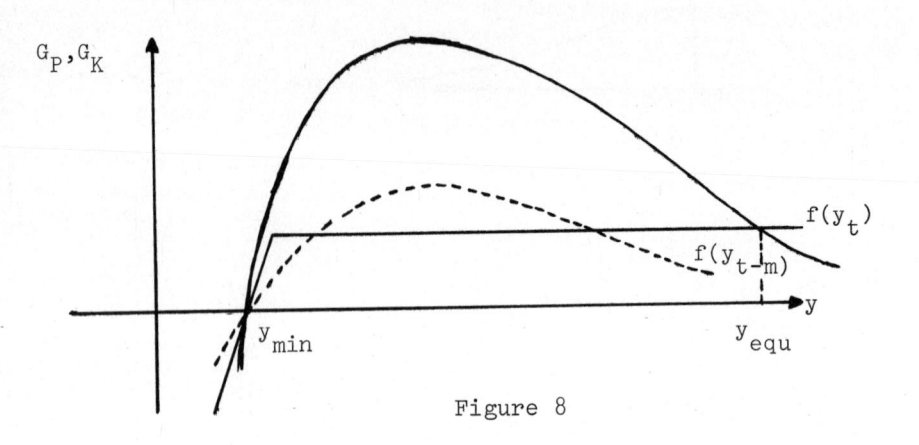

Figure 8

In the long run this model delivers the well known neoclassical equilibrium solution with a constant population growth rate. The population grows at the rate $G_P = G_{P_{max}}$

$$G_Y = \frac{\dot{Y}}{Y} = G_K = \frac{\lambda}{1-\alpha} + G_P$$

The per-capita-income grows at the rate

$$G_y = \frac{\dot{y}}{y} = \frac{\lambda}{1-\alpha}$$

The stability of this equilibrium solution can be demonstrated: From equation (38) we get

$$G_{K_{t-1}} = s \cdot 1^{\frac{1-\alpha}{\alpha}} \cdot e^{\frac{\lambda}{\alpha}(t-1)} \cdot a^{\frac{1}{\alpha}} [y_{t-1}^{\frac{\alpha-1}{\alpha}} - y_{min} \cdot y_{t-1}^{-\frac{1}{\alpha}}] \quad (39)$$

and

$$G_{K_t} = s \cdot 1^{\frac{1-\alpha}{\alpha}} \cdot e^{\frac{\lambda}{\alpha}t} \cdot a^{\frac{1}{\alpha}} [y_{t-1}^{\frac{\alpha-1}{\alpha}} \cdot e^{G_{y_t}(\frac{\alpha-1}{\alpha})} - y_{min} \cdot y_{t-1}^{-\frac{1}{\alpha}} \cdot e^{-\frac{1}{\alpha}G_{y_t}}]$$

(40)

If we substitute for G_{y_t} the expression $\frac{u_t}{1-\alpha}$, where u_t is an auxiliary variable and $u_t = \lambda$ if G_{y_t} is equal to the equilibrium growth rate $\bar{G}_{y_t}$ and $u_t \gtreqless \lambda$ if $G_{y_t} \gtreqless \bar{G}_{y_t}$, we get by dividing (40) by (39).

$$\frac{G_{K_t}}{G_{K_{t-1}}} = \Phi_t \cdot e^{\frac{\lambda - u_t}{\alpha}} \qquad (41)$$

where

$$\lim_{y_{t-1} \to \infty} \Phi_t = \lim_{y_{t-1} \to \infty} \frac{1 - \frac{y_{min}}{y_{t-1}} \cdot e^{\frac{-u_t}{1-\alpha}}}{1 - \frac{y_{min}}{y_{t-1}}} = 1$$

If we now have:

$G_{y_t} < \bar{G}_y$ or $u_t < \lambda$ then $e^{\frac{\lambda - u_t}{\alpha}} > 1$ which means $G_{K_t} > G_{K_{t-1}}$. If the G_K increases, G_{y_t} will also increase and reach $\bar{G}_y$. If we start with a growth rate of the per-capita-income greater than the equilibrium growth rate: $G_{y_t} > \bar{G}_y$ or $u_t > \lambda$ then $e^{\frac{\lambda - u_t}{\alpha}} < 1$ and $G_{K_t} < G_{K_{t-1}}$. If the growth rate of capital decreases, the growth rate of the per-capita-income will decrease as well and finally reach the equilibrium rate $\bar{G}_y$.

In order to take account of increasing negative externalities connected with a growing population we introduce analogous to our function (5) a function g

$$g = g(P) \qquad (42)$$

with the properties

$$g(o) = 1, g(P^x) = 0$$

$$\frac{\partial g}{\partial P} \leq 0, \frac{\partial^2 g}{\partial P^2} < 0,$$

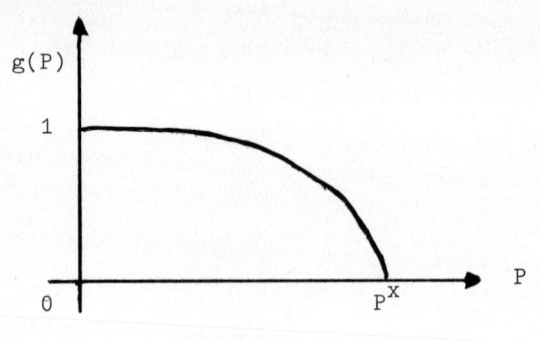

[Figure 9: graph of g(P) vs P, showing g(P)=1 for low P, decreasing to 0 at P^X]

Figure 9

where P^X is defined analogous to L^X in equation (8).

The production function modified by (12) is now :

$$Y = a\, K^\alpha\, L^{1-\alpha} \cdot e^{\lambda t} \cdot g(P) \qquad (32')$$

Thus the multiplication with g(P) has a similar effect of reduction of the value of production as the substraction of negative external effects by equation (5).

Equation (32') leads to

$$y = 1\, a\!\left(\tfrac{K}{L}\right) \cdot e^{\lambda t} \cdot g(P) \qquad (37')$$

and

$$G_K = s \cdot 1^{\tfrac{1-\alpha}{\alpha}}\, \left(\tfrac{a}{y}\right)^{\tfrac{1}{\alpha}} \cdot e^{\tfrac{\lambda}{\alpha} t}\, (y - y_{min})\, g(P) \qquad (38')$$

We will now discuss how the function (38') will be shifted over time. This will also answer us the question if the equilibrium growth rate of population can stay constant if we take into account negative externalities.

At relatively low population densities g(P) is still equal to one. The economy grows according to the neoclassical equilibrium described above :

$$G_P = G_{P_{max}}$$

and

$$G_K = G_Y = \frac{\lambda}{1-\alpha} + G_P$$

This upward shift of the function (38') by technical progress is slowed down and finally compensated by g(P) if the population size increases. Obviously, this is the case as long as G_P is positive. Graphically this can be demonstrated as follows:

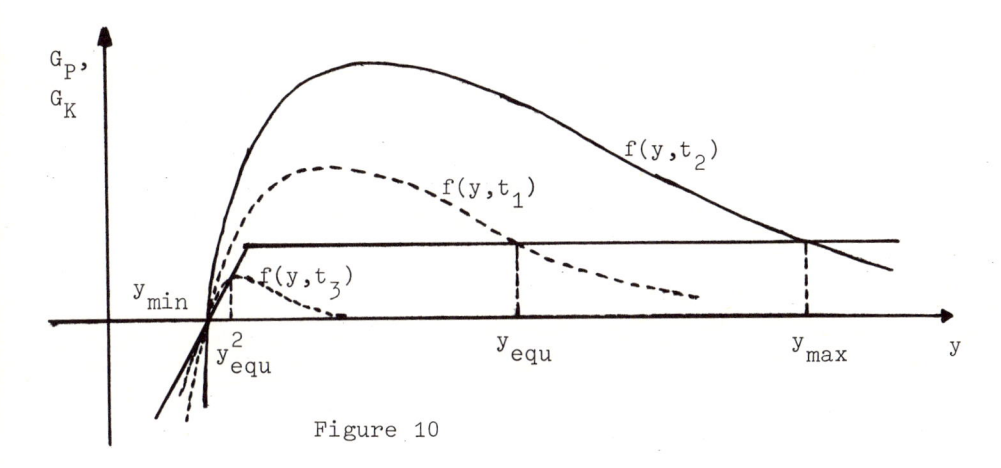

Figure 10

The graph shows three different functions $G_K = f(y, t, P)$ at three different time periods. With an increasing population the function will be shifted by technical progress from $f(g, t_1)$ up to $f(g, t_2)$. Further increases will approach the function more and more to the y-axis. This implies the existence of a maximum of the per-capita-income. It also follows that y will approach y_{min} (the income at which the population is constant and net saving is zero) and G_P and G_K will approach zero.

The so-called classical population theory does not allow to stop the population growth at the point where y is at its maximum. In other words, the population will not stay at the size which we have defined above as optimal.

The classical population theory has been criticized for attempting to explain human behavior by a general observed phenomenon of animal behavior. Other theories have been proposed, but none of them has gained general recognition (Leibenstein, 1957; Hagen, 1959; Krelle, 1965). A more realistic theroy has to explain why the birth rate b declined in many high-income countries. From our model we can see that this leads to a lower G_P and G_K, but to a higher per-capita-income. Only if the population theory would yield a zero growth rate of population before y has gone back to y_{min} we could have a stable population at an income level higher than y_{min}.

Even if we could find amore realistic population theory an equilibrium at the maximum of y is very unlikely. This will particularly be the case because the costs of negative external effects are never fully taken into account so that the economy might even have the illusion of constantly increasing per-capita-income while "net" per-capita-income is falling rapidly.

V. Conclusion

In the first part of this paper we have examined the static population optimum. In order to define the "net" production function we used the calssical production function together with a function of negative external effects. The optimum size of the population was defined as the point where "net" per-capita income reaches its maximum. It became obvious that the population optimum (L^{opt}) is always smaller than the maximum of the "net" production function ($\underline{L}$).

Each economy has the alternative to allocate its capital to production or to a reduction of negative externalities. An economy with an optimum size of population allocates in an optimum way if the value of the "net" marginal product of capital in production is equal to the value of the marginal product resulting from a reduction of negative external effects.

Only if we know the type of technical progress and the parameters or parameter changes we are able to make a clear statement about the effects of technical progress on the population optimum. The well-known Hicks-neutral technical progress does not change the population optimum.

In a model of neoclassical economic growth with endogenous population growth we demonstrated the consequences of negative external effects. Independent from the population theory - classical or any other realistic theory - we have found it very unlikely that an economy will stop its population growth at the maximum of the per-capita-income.

Any conclusion for a population policy will only be possible if we estimate the function of negative externalities (G) and the production (F). If we find that the slope of the function G in the relevant area is greater than the slope of the production function we know that we are already beyond the maximum of the "net" production function, which means that we have gone beyond the optimum size of population.

References

Arrow, K.J., 1951, Social Choise and Individual Values, Cowles Foundation Monograph 12 (John Wiley, New York).

Cannan, E., 1903, Elementary Political Economy, 3rd edition, London.

Hagen, E.E., 1959, Population and Economic Growth, The American Economic Review, 49.

Krelle, W., 1965, Beeinflussbarkeit und Grenzen des Wirtschaftswachstums, Jahrbücher für Nationalökonomie und Statistik, 178.

Leibenstein, H., 1957, Economic Backwardness and Economic Growth, Studies in the Theory of Economic Growth, New York and London.

Nelson, R., 1956, The Low-Level-Equilibrium Trap in Underdeveloped Economies, The American Economic Review, 46.

Ohlin, G., 1967, Population Control and Economic Development, (OECD, Paris).

Penrose, E.F., 1934, Population Theories and their Application, Stanford.

Phelps, E.S., 1967, Golden Rules of Economic Growth (North-Holland, Amsterdam).

Pitchford, J.D., 1974, Population in Economic Growth,(North-Holland, Amsterdam).

Russel, R.R., 1972, The Optimum Population and Growth : A Comment, Journal of Economic Theory, 5.

Sauvy, A., 1956, Théorie générale de la population, Paris.

Votey, H.L., 1969, The Optimum Population and Economic Growth : A New Look. Journal of Economic Theory, 1.

Wicksell, K., 1910, Das Optimum der Bevölkerung, in Die neue Generation, 6, pp. 383-391.

SPATIAL EQUILIBRIUM IN THE DISPERSED CITY

Martin J. Beckmann[x]
Technische Universität München, W.Germany and
Brown University, U.S.A.

I. Introduction

The standard model of residential land use in a city as treated by Mills (1973) assumes that all working and shopping opportunities are concentrated in the center of the city, the Central Business District (CBD) and that residential land is homogenous otherwise. Through this assumption, a definite orientation with respect to the center is introduced into residential land use. The effect of any other interactions over distance is overlooked. To put it bluntly, work and consumption (shopping) dominate all trip making behaviour, the interaction with other residents through social and recreational contacts is completely ignored.

In this paper we will focus instead on interaction among households, on the utility of this interaction to the individual household, and on the role of this interaction in shaping equilibrium patterns of residential living. For simplicity and transparency we shall ignore work and shopping trips completely. An alternative interpretation would be to assume that work and shopping have the same areal distribution as population.

In order to keep things mathematically simple we assume a one dimensional - a long narrow city - in the sense of Solow and Vickrey (1971). The amount of land allocated to transportation is not treated explicitly. It is assumed that transportation cost, or more precisely the time cost of interaction for any pair of households is proportional to their distance. The utility function for a household is assumed to depend on two variables : the cost of interaction with others as measured by the average distance of this household from all other households in

---

[x] The assistance of J. Fisher in the calculations and drawing of figures is gratefully acknowledged.

the city, and the amount of space occupied. In addition there may be
a utility of consumption but this turns out to have no effect on spatial
behaviour, if the utility function is separable with respect to the
three variables. Specifically utility is treated as logarithmic with
respect to space and linear with respect to average distance and with
respect to other consumption.

$$u(x) = a \log s(x) - \bar{r}(x) + c(x) \qquad (1)$$

where
u = utility, s = space, $\bar{r}$ = average distance
c = consumption
a = a parameter

and x denotes the location of a household.

II. The Model

We assumed free mobility : each household is free to choose among
locations it desires. We observe first that at given prices $p(x)$ a
household will choose the same amount of space regardless of income.
For given income y and price $p(x)$ the household's utility becomes

$$u(x) = a \log s(x) - \bar{r}(x) + y(x) - p(x)s(x) \qquad (2)$$

Maximization of (2) with respect to s yields

$$\frac{a}{s(x)} - p(x) = 0 \qquad \text{or}$$

$$s(x) = \frac{a}{p(x)} \qquad (3)$$

Expenditures on living space is thus constant and equals a and (2)
becomes

$$u(x) = a \log s(x) - \bar{r}(x) + y(x) - a \qquad (4)$$

where $s(x)$ is determined by (3) from the given $p(x)$. Utility is thus
a linear function of income.

While this is a special property of the logarithmic utility function, the same result would apply if we assume that all households pay the same rent (possibly zero) and that the amount of space $s(x)$ allocated to every household must be so such that everybody is satisfied, i.e. indifferent among locations.

A necessary condition for equilibrium is therefore that utility, given y, is constant, i.e.

$$a \log s(x) - \bar{r}(x) = \text{constant} \tag{5}$$

In terms of density $m(x)$ where

$$m(x) = \frac{1}{s(x)}$$

the condition (5) becomes

$$a \log m(x) + \bar{r}(x) = \text{constant} = u_o \tag{6}$$

We shall now establish a relationship between the density distribution $m(x)$ and the average distance function $\bar{r}(x)$. Assume that the city extends between points $-R$ and R along the x-axis. By definition

$$\bar{r}(x) = \int_{-R}^{x} m(r)(x-r)dr + \int_{x}^{R} m(r)(r-x)dr \tag{7}$$

Differentiation yields

$$\frac{d^2 \bar{r}(x)}{dx^2} = 2m(x) \tag{8}$$

Although we are mainly interested in $m(x)$ it will be convenient to consider $\bar{r}$ first. From (6) and (8) it follows:

$$a \log \tfrac{1}{2} \bar{r}(x) = u_o$$

or

$$\bar{r}(x) = 2 e^{\frac{u_o - \bar{r}}{a}} \tag{9}$$

By a suitable choice of units for time and distance

$$v = \frac{\bar{r}}{a}$$

$$\zeta = 2x(\frac{1}{a} e^{-\frac{u_o}{a}})^{-\frac{1}{2}}$$

Equation (9) may be standardized into :

$$v'' = \frac{1}{2} e^{-v}$$

where

$$v'' = \frac{d^2 v}{dx^2}$$

Its range of definition is the positive axis $\zeta \geq 0$.

III. Solution of the Model

A solution to equation (10) in closed form may be found as follows. Multiply (9) by $2v'$ and integrate

$$\int 2v'v'' dx = \int e^{-v} v' dx$$

where we have written x again for ζ.

Now

$$(v')^2 = -e^{-v} + c_1 \quad \text{or}$$

$$\frac{dv}{\sqrt{c_1 - e^{-v}}} = dx \qquad (11)$$

Let

$$c_1 - e^{-v} = z^2 \qquad (12)$$

$$e^{-v} dv = 2z\, dz$$

$$dv = \frac{2z\, dz}{c_1 - z^2}$$

Substituting in (11) we get

$$\frac{2\, dz}{c_1 - z^2} = dx \qquad (13)$$

Let
$$c_1 = k^2 \text{ and observing that}$$

$$\frac{2}{k^2 - z^2} = \frac{1}{k}\frac{1}{k+z} + \frac{1}{k}\frac{1}{k-z} \text{ we obtain from (13) by substituting}$$

and integrating

$$\int \frac{1}{k}\frac{dz}{k+z} + \int \frac{1}{k}\frac{dz}{k-z} = \int dx$$

$$\frac{1}{k}\log\frac{k+z}{k-z} = x + c_2$$

$$z = k\frac{e^{k(x+c_2)} - 1}{e^{k(x+c_2)} + 1} \qquad (14)$$

Recall the definitions (12) of z and (13) of k.

$$k^2 - e^{-v} = k^2\left(\frac{e^{k(x+c_2)} - 1}{e^{k(x+c_2)} + 1}\right)^2$$

From this it follows

$$v = -\log k^2\left[1 - \left(\frac{e^{k(x+c_2)} - 1}{e^{k(x+c_2)} + 1}\right)^2\right] \qquad (15)$$

In order that the solution be valid on the negative axis we must set

$$v = -\log k^2\left[1 - \left(\frac{e^{k(|x| + c_2)} - 1}{e^{k(|x| + c_2)} + 1}\right)^2\right]$$

In order to determine c_2 observe that average transportation time has a minimum at $|x| = 0$. For this it is necessary and sufficient that $c_2 = 0$. Thus finally

$$v(x) = -\log k^2 \frac{4 e^{k|x|}}{(1+e^{k|x|})^2} \qquad (16)$$

The function of interest, housing density $m(x)$, may now be found by substituting (16) in the differential equation (11):

$$m(x) = \frac{1}{2} v''$$
$$= \frac{1}{4} e^{-v}$$

or

$$m(x) = k^2 \frac{e^{k|x|}}{(1+e^{k|x|})^2} \tag{17}$$

The unknown coefficient k is to be determined by the condition that a given number of households N be housed in the given space 2R.

$$N = 2 \int_0^R \frac{1}{m} dx$$
$$= \frac{2}{k^2} \int_0^R \frac{(1+e^{kx})^2}{e^{kx}} dx$$
$$= \frac{2}{k^3} e^{kR} + \frac{4R}{k^2} - \frac{2}{k^3} e^{-kR} \tag{18}$$

Figure 1 shows how k depends on N for given R = 1. The function (17) which except for a factor k may be interpreted as the derivative of the logistic function $\frac{1}{1+e^{-kx}}$ is decreasing from $m(o) = \frac{k^2}{4}$ to arbitrarely small values as $|x|$ is increased.

The <u>conclusions</u> to be drawn from the solution (17) are as follows. In the dispersed city, locations continue to differ with respect to centrality. Here central locations enjoy greater proximity and hence better opportunities for contacts with others. To compensate for this households are squeezed into more compressed quarters.

The competitive equilibrium of this market is <u>Pareto optimal</u>. The greatest utility sum is squeezed out of a given amount 2R of space for a given number N of households by inducing a density distribution which is <u>symmetric</u> and <u>decreasing</u> from the center. The distribution is such that every individual is indifferent among locations, space being traded for proximity to others at rates which are equally acceptable to all. General economic principles suggest that this also yields the greatest sum of utility. (A formal proof in terms of a calculus of variations problem is not difficult).

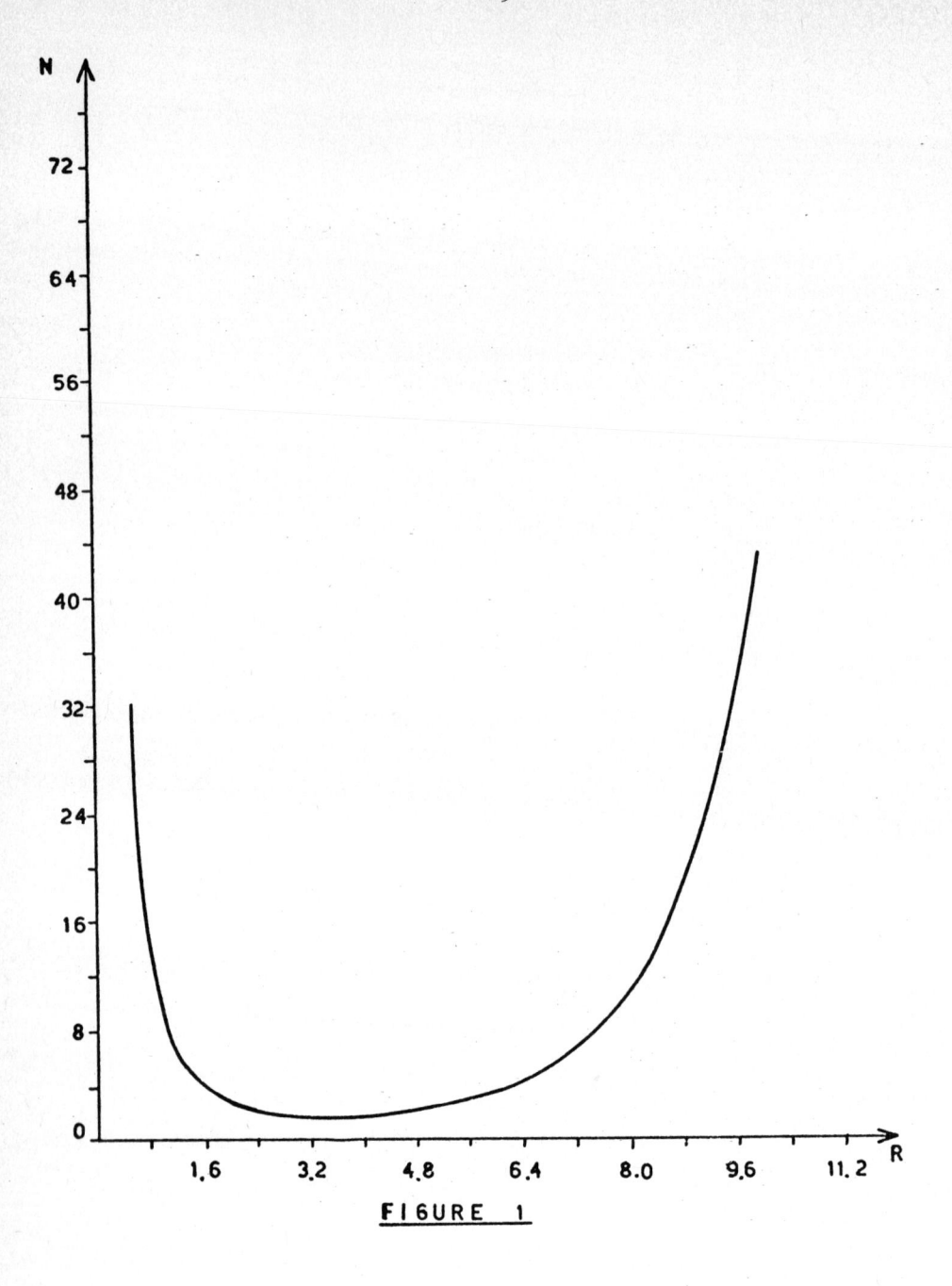

FIGURE 1

IV. Comparison of a Dispersed Equilibrium with a C.B.D. Model

Compare now the dispersed equilibrium with that of a city with dominant DBD. In the latter case in equation (5)

$$a \log s(x) = \text{constant} + |x| \qquad (19)$$

$$s(x) = e^{c_3 + \frac{|x|}{a}}$$

Since we have standardized $a = 1$ the density distribution is

$$\tilde{m}(x) = m(o) \, e^{-|x|} \qquad (20)$$

For comparable boundary conditions in equation (17), a k value must be found which results from the same area 2R and the same population N. As an example consider $R = 1$ and $k = 1$. The corresponding N-value may be calculated as

$$N^{\star} = 4 + 2e - \frac{2}{e} \qquad (21)$$

$$= 8.700805$$

The value of $m(o)$ is then obtained from

$$N^{\star} = \int_{-R}^{R} m(x) \, dx = 2m(o) \int_{0}^{1} e^{-x} dx \qquad (22)$$

hence

$$m(o) = \frac{N^{\star}}{2(1 - \frac{1}{e})} \qquad (23)$$

$$= 1.850918$$

The graphs of functions (17) and (20) for this special case have been plotted in Figure 2. It is interesting that in the case of the dispersed city the density graph is smooth at the center, whereas in the concentrated case it has a sharp peak.

It is of interest also to study the savings in transportation cost that are achieved by the relative compressions of density in the central parts of the city in the decentralized case. If N families are housed at a uniform density δ in the space 2R, then the resulting uniform

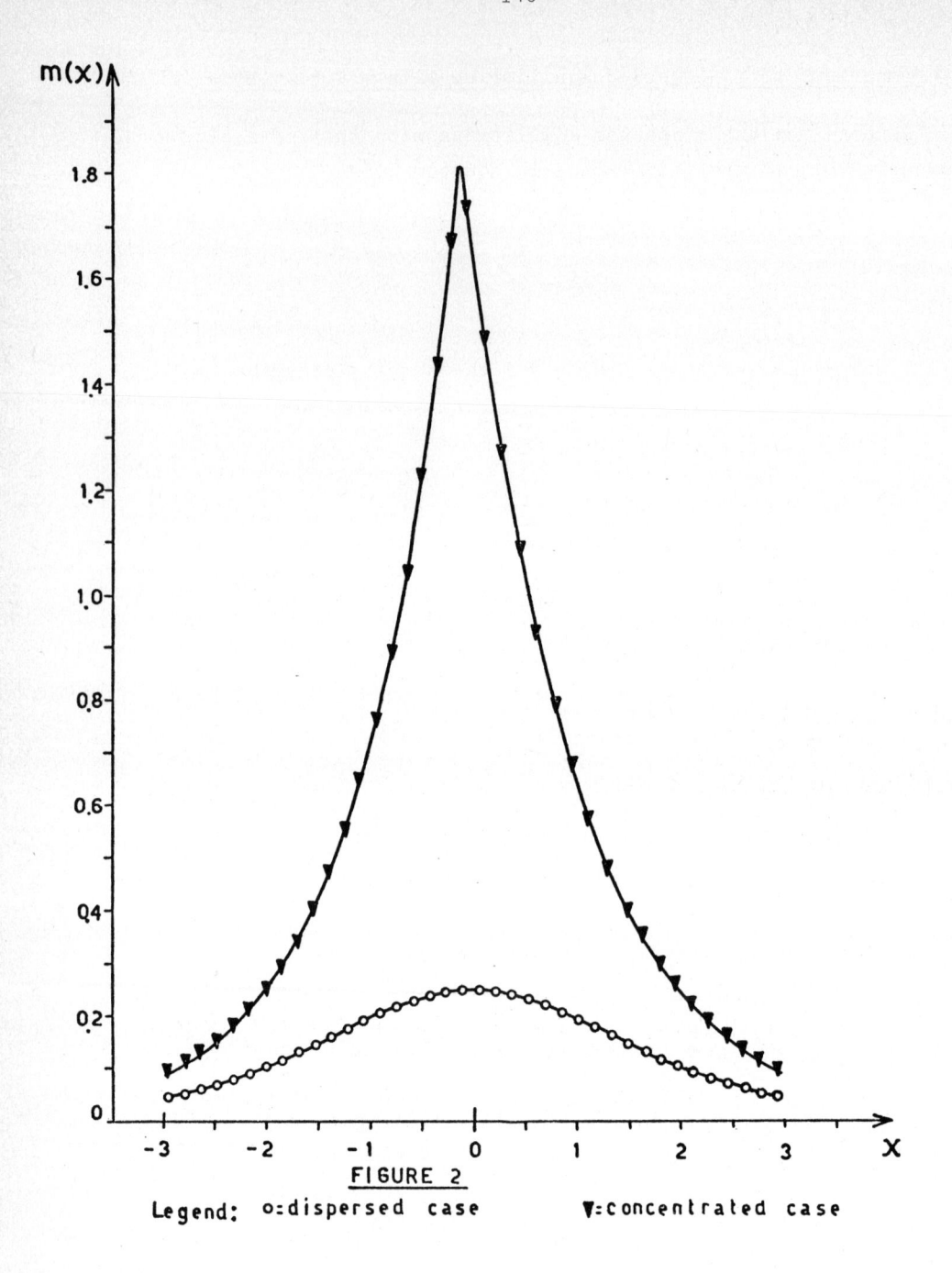

FIGURE 2

Legend: o=dispersed case   ▼=concentrated case

density is $\frac{N}{2R}$ and the resulting total time cost of transportation $\bar{r}$ is

$$\bar{\bar{r}} = \delta \int_{-R}^{R} \bar{r}(x)dx = 2\delta \int_{o}^{R} \bar{r}(x)dx$$

$$= 2\delta \int_{o}^{R} [\int_{-R}^{x} m(r)(x-r)dr + \int_{x}^{R} m(r)(r-x)dr]\, dx$$

$$= \frac{N}{R} \int_{o}^{R} [\int_{-R}^{x} (x-r)dr + \int_{x}^{R} (r-x)dr]\, dx \qquad (24)$$

or

$$\bar{\bar{r}} = \frac{1}{3} NR^2$$

In the case $R = 1$, $k = 1$, $N = N^*$ one has

$$\bar{\bar{r}} = \frac{1}{3} N^* \approx 2.900268 \qquad (25)$$

In the dispersed case we have

$$\bar{\bar{r}} = 2 \int_{o}^{1} [\int_{-1}^{x} \frac{e^r(x-r)}{(1+e^r)^2} dr + \int_{x}^{1} \frac{e^r(r-x)}{(1+e^r)^2} dr]\, dx = 0{,}85 \qquad (26)$$

It appears from this example that the density distribution in the dispersed case achieves a saving of 71 percent on the total time cost of transportation as compared to the city model with a dominant C.B.D.

References

Beckman, M.J., 1969, "On the Equilibrium Distribution of Urban Rent and Residential Density", Journal of Economic Theory, 1, no.1, 60-67.

Mills, E.S. and Mc Kimmon, J., 1973, "Notes on the Urban Economics", Bell Journal of Economics and Management Science, 4, no. 2, 539-601.

Solow, R. and Vickrey, W., 1971, "Land Use in a Long Narrow City," Journal of Economic Theory, 3, no. 4, 430-447.

A RATIONALE FOR AN URBAN SYSTEMS MODEL (USM)

Christopher G. Turner
Nathaniel Lichfield and Associates
London - England

I. Theoretical Foundation

The urban system model (henceforth denoted by USM) is based upon a conceptualization of the urban system as the culmination of a process by which physical stock (e.g. translation and employment) are distributed across the metropolitan region (Echenique, 1968). The hypothesis here is that activities are distributed to locations as a function of their interrelationships with other activities and the constraints imposed by the physical stock; and physical stock locates in response to the activity demands for stocks - for example, for transportation, space and infrastructure. The model also explicitly incorporates the competition for physical stock between urban activities through an accounting framework which relates the distribution of activities to the availability of physical stock.

Specifically, the USM distinguishes between "growth generating" (primary) employment, residential population, and service activities; and between floorspace, transportation and public utility stocks. The model simulates changes in the distribution of primary employment, residential population and service employment over time as a function of changes in intrinsic locational attractiveness and changes in the availability and quality of the physical stock.

The model is founded principally on the following theoretical contributions: the work carried out by Lowry for the Pittsburgh Comprehensive Renewal Program; research into entropy maximizing concepts at the Centre for Environmental Studies, London; urban land market theory, and central place theory. In order to understand the way in which the structure of the USM has been established in is necessary to consider each contribution in turn. The significance of Lowry's contribution (Lowry, 1964) is that he formalized and tested a stratification of metropolitan economic activity in terms of a basic (export oriented) residential population and a service (population serving) sector. These were interrelated globally through a series of activity and population serving ratios and spatially through a series of zonal activity distribution functions.

However, the definition of the metropolitan economy in Lowry's terms leads to a number of inconsistencies, the most important of which is the fact that all economic growth is assumed to be export oriented. No allowance is made for growth through either import saving or local market oriented employment. To overcome this problem the USM recognises that both import saving and certain local market oriented economic activities, such as local government expenditure or investment in the housing market, could increase the regional economic growth rate (Tiebout, 1962). It therefore differentiates between those economic activities which are considered capable of generating economic growth and those activities which are considered to be ultimately dependent upon the regional market.

Growth generating (primary) activity is defined in terms of the following employment categories: export oriented, unique locating, import saving, and certain local market oriented activities which influence the overall study area rate of economic growth. The service sector is defined in terms of the remaining local market serving activities. What is more, unlike the Lowry model and a number of its descendents (see for example Goldner, 1971) the USM explicitly incorporates a procedure to predict the future zonal distribution of primary employment.

The second important influence on the development of the USM has been the research carried out at the Centre for Environmental Studies, London. Perhaps the most important contribution of this work has been the extablishment of a general theoretical framework for the analysis of urban activity distribution (Wilson, 1967, 1970).

This framework is based on the concept of entropy as used in statistical mechanics, and relies upon a definition of the micro and macro properties which characterise the system of interest. Assuming as an example the journey-to work system, which is concerned with distributing individuals from residential origin locations to workplace destination locations, the micro properties of that system can be defined in terms of specific individual movements between origins and destinations which satisfy any constraints on movement in the system. The macro property of the system is defined in terms of a distribution of movements irrespective of individual movements. As numerous individual movements in the system can form any one distribution, and assuming that all individual movements are equiprobable, the theory is based on the derivation of the most probable distribution of person movements which satisfies any constraints imposed on the system. Wilson proves that the most probable distribution of person movements can be derived by maximizing the entropy of the system.

In terms of the USM, this theoretical derivation is all-important for the development of the residential and service sub-models. The spatial distribution of residential and service activities is determined by sub-models which respectively distribute employees from work-to-home, and service demand from home- and work-to-service centres, and in so doing maximizes the entropy of each activity systems.

Considering now urban land market and central place theory, it is significant that the residential sector trade off decision between transportation "cost" and locational attractiveness, predicted in the form of a spatial distribution of population, is indirectly related to the urban land market theories of Wingo (Wingo, 1961) and Alonso (Alonso, 1964). Similarly, the service sector trade off decision, predicted in the form of a spatial distribution of service centres, has much in common with the spatial hierarchy of service centres suggested by Central Place Theory (Berry, 1967).

The conclusion of Wingo and Alonso is essentially that both residential land value and residential density decline as the "cost" of transportation increases away from employment centres. The decay of land value and population density is representative of the trade off decision made by households or individuals between the "cost" of commuting and the attractiveness of residential locations. Thus, those individuals or households having higher space preferences tend to locate further away from their place of work, as the unit price of space declines.

Central place theory is very relevant to the service sector in that the predicted distribution of services simulates the basic central place concepts, and specifically the hypotheses that:

(1) centres with superior centrality (accessibility) provide a greater range of goods and services at a higher level of production than less accessible centres;
(2) consumers are willing to travel farther, in terms of the transportation "cost" incurred in order to reach the more accessible centres, which provide a wide choice of goods and services, than they are to smaller and less accessible centres;
(3) smaller centres perform only limited scale activities and serve limited market areas, whereas larger centres perform a wider range of activities and attract consumers from a more extensive market area.

The theoretical contributions described above have all served to influence the basic structure of the USM activity distribution algorithm.

In addition a series of efficiency and equity impact prediction techniques have been explored so as to make the USM a more useful metropolitan planning tool. These are considered in the following section.

II. Model Structure

The starting point is the assumption that the total amount and distribution of residential population and service employment is determined by the total amount and sitribution of regional "growth generating" (primary) employment within the region, which is distributed to zones through the primary employment sub-model

$$P^m_{i_{t1}} = f(I_{i_{t0}}, E_{i_{t0}}, AC_{i_{t1}}, DP_{i_{t0}}, N_{i_{t0}} \ldots) \qquad (1)$$

where $P^m_{i_{t1}}$ = type m primary employment located in zone i at time t1

$I_{i_{t0}}$ = lagged inter industry characteristics (e.g. zonal intensity of base year service sector activity

$E_{i_{t0}}$ = lagged economies of scale characteristics (e.g. zonal intensity of base year primary sector activity in zone i)

$AC_{i_{t1}}$ = accessibility characteristics (e.g. base year zonal accessibility to airports, freeway interchanges, rail heads, docks)

$DP_{i_{t0}}$ = lagged development potential characteristics (e.g. base year zonal amount of vacant industrial land)

$N_{i_{t0}}$ = lagged non-compatible use characteristics (e.g. base year zonal intensity of residential development).

These employees are then distributed from workplace to residential locations through the residential sub-model and an activity rate is applied to primary employees now located in residence to predict the primary employee dependent population.

The primary employees and their dependents set up a demand for services which is determined from a population and employment serving ratio and distributed to metropolitan service centres by the service sub-model. In this way the number of service employees necessary to satisfy consumer demand at each service centre is predicted.

Service employees themselves establish a demand for residences which is distributed from workplace to residential locations by the residential sub-model, with the activity rate again applied to determine the incremental amount of population to be added to the previously located population. The service demand set up by this increment of population and the previous increment of (service) employment is then predicted, distributed to service centres by the service sub-model, and a new incremental amount of service employment is predicted at each service centre.

This procedure is reiterated with diminishing incremental amounts of population and service employment being generated and distributed across the region at each iteration.

At this point also, the predicted zonal residential population and service employment activity distributions are checked against the available physical stock of space which is input to the Model in the form of zonal holding capacities. The predicted residential population is redistributed by the residential sub-model so that the location of population is in balance with the constraints imposed on location by the available stock of residential space (holding capacity).

At this point the distribution of the service employment is out of balance with both the final distribution of service demand (i.e. population and employment), and also with the constraints imposed on location by the available stock of service space at each location. The service sub-model is therefore reiterated and the distribution of service employment is brought into balance with both the final service demand distribution and with the available stock of service space (holding capacity).

The general form of the unconstrained and constrained residential and service sector sub-models can be expressed as follows:

$$T_{ij_{t1}} = A_{i_{t1}} O_{i_{t1}} B_{j_{t1}} D_{j_{t1}} \exp(-\beta c_{ij_{t1}}) \quad (2)$$

with

$$A_{i_{t1}} = [\Sigma_j B_{j_{t1}} D_{j_{to}} \exp(-\beta c_{ij_{t1}})]^{-1} \quad (3)$$

and

$$B_{j_{t1}} = [\Sigma_i A_{i_{t1}} O_{i_{t1}} \exp(-\beta c_{ij_{t1}})]^{-1}$$

for the constrained sub-model

$$B_{j_{t1}} = 1 \quad (4)$$

for the unconstrained sub-model

subject to :

$$\sum_j T_{ij_{t1}} = O_{i_{t1}} \quad (5)$$

$$\sum_j T_{ij_{t1}} = P^*_{j_{t1}} \quad (6)$$

$$\sum_{ij}\sum T_{ij_{t1}} c_{ij_{t1}} \left[\sum_{ij}\sum T_{ij_{t1}} \right]^{-1} = \overline{C}_{t1} \quad (7)$$

where $T_{ij_{t1}}$ = flow of workers or service demand between zone i and j at time t1

$O_{i_{t1}}$ = number of residential or service sector flow origins in zone i and time t1

$P^*_{j_{t1}}$ = number of residential or service sector flow destinations in zone j at time t1

$c_{ij_{t1}}$ = travel cost between zone i and j at time t1

β = travel cost parameter

$\overline{C}_{t1}$ = mean regional cost of travel in residential or service sector at time t1

A_i, B_j = defined by equations (3) and (4).

Once the distributions of population and employment are in balance with the constraints, the travel characteristics associated with them are estimated.

To this point in the process the model only provides a description of future urban conditions in terms of activity distributions and travel characteristics under different sets of policy or forecasting assumptions and gives little if any indication as to the relative desirability of any one policy versus another.

For this reason a series of sub-models are being developed to transform the descriptive information output by the USM distribution sub-models into efficiency and equity measures that can be used in the comparative assessment of policy alternatives. These are as follows.

III. Efficiency Measures

1. A measure of net travel benefit incorporating user cost and trip end benefits (Neuburger, 1971). For negative exponential distribution

sub-models such as those incorporated in the USM this can be expressed as :

$$S = \sum_i PP_i \{\frac{1}{\beta}[\log(\sum_j W_j^2 e^{-\beta c_{ij}^2}) - \log(\sum_j W_j^1 e^{-c_{ij}^1})] \quad (8)$$

where S   = consumer surplus of plan 2 over plan 1
       PP_i = no of people (e.g. residents, employees, shoppers) in zone i
       W_j = zonal attractiveness of zone i
       c_{ij} = cost of travel between i and j
       β = coefficient on the cost of travel.

The importance of this development is that it explicitly considers the effect of variations in both land use and transportation on net travel benefits, and it can also be derived from the USM without recourse to the conventional transportation model procedure.

2. A measure of the degree of consistency between the policy assumptions being tested. What is important here is the rate at which the model converges on the pattern of development being tested, which can vary considerably depending on the consistency of the development and transportation assumptions and the difficulty which the model has in simulating different policies, for example, as between two land use alternatives with the energy pricing assumptions. A statistical measure of the rate of convergence is therefore being developed as one indicator of the potential "fit" between the transportation and development policies being tested. The more rapid the convergence the better the fit, and vice versa.

In common with most models of this type, the USM incorporates an optional zonal holding capacity procedure which can be applied to the distribution of both population and employment, and through the manipulation of which, alternative urban development policy assumptions can be input and tested by the model. Now, as currently formulated, these capacities do not represent absolute constraints and can be overridden - if, for example, an individual zone is highly accessible but has an unrealistically low development ceiling. This flexibility can be seen from the general form of the constraints procedure given in equations (9), (10), (11) and (12) respectively, where, if :

$$D_j/D_j^\star < 1, \; j\epsilon z_1 \quad (9)$$

$$D_j/D_j^\star > 1, \; j\epsilon z_2 \quad (10)$$

then :

$$B_j(m) = \begin{cases} B_j(m-1)D_j/D_j^\star, & j\varepsilon z_1 \\ 1, & j\varepsilon z_2 \end{cases} \quad (11)$$

with : D_j = population or service employment holding capacity of zone i

$D_j^\star$ = predicted population or service employment of zone i

z_1 = variable subset of constrained zones

z_2 = variable subset of unconstrained zones

m = iteration number of population or service employment constraints procedure

B_j = value by which the relative attractiveness of zone j as a location is reduced.

The relevant values of B(m) are then substituted into the residential or service distribution sub-model respectively which is reiterated until :

$$D_j/D_j^\star \geq j\Sigma z_1 \quad (12)$$

3. A measure of the efficiency of alternative public utility policies. Currently, an empirical relationship verified in Detroit by Doxiadis (1967) is being applied to the population density gradients estimated by the USM to derive a measure of the differential cost of providing public service (sewer and water) utilities. This relationship, which takes the form of a parabolic curve, can be expressed as :

$$U_{(i)} = \sqrt{23333136 - [2304((PD_{(i)}-100^2)]/10609))} - 7.0 \quad (13)$$

where : $U_{(i)}$ = $ cost of public service utilities/square mile

$PD_{(i)}$ = net population density (thousands of residents/square mile).

4. A measure of mobile source air pollution. Recent Federal legislation, particularly the National Environmental Policy and Clean Air Acts of 1973, has served to focus considerable attention upon the achievement and maintenance of air quality standards (see for example, Bellomo, 1972).

A simple sub-model has therefore been developed to assess the impact of alternative policies on the emission of atmospheric pollutants resulting from the transportation system. Although the ultimate concern is with reducing the concentrations of pollutants in the atmosphere, United States Federal legislation also requires the development of control strategies by which aggregate pollutant emissions can be reduced to a level consistent with the achievement and maintenance of a national standard.

The air pollution sub-model incorporated in the USM utilises vehicle mile emission factors for carbon monoxide, hydrocarbons, nitrogen oxides, sulphur oxides and particulates respectively.

5. A measure of energy consumption based on an estimate of BTU's expended under each policy alternative, with the required vehicle miles of travel and speed assumption being generated from a simple model such as that developed by Koppelman (1969) and verified in New York and Washington D.C.. This can be expressed as :

$$VMT = 64.3 \ VTO^{-74} e \ (1.6 \ FE/FO) \quad (14)$$

where : VMT = vehicle miles of travel/square mile
       VTO   = vehicle trip origins/square mile
       FE/FO = proportion of total roadway surface area made up by freeways.

IV. Social Equity Measures

By their very nature models such as the USM are restricted in the range of information which they can provide for assessing the comparative social equity of policy alternatives. Perhaps the most commonly used and often abused area to be explored is that of accessibility. Alternative measures of accessibility to opportunities such as jobs, shops and homes have been used in many studies to provide an assessment of the relative advantages of one policy against another. Weaknesses in the use of such measures have manifested themselves in two main ways, firstly in the form of measures used (Whitbread, 1972) and secondly in their general application as aggregate measures of plan advantage (Turner, 1972).

In view of these problems the approach taken in the development of the USM has been to identify for different socio-economic groups the numbers of specific opportunities of different types - job types, dwelling types, cultural, recreational, social and medical facilities - available to them within specified ranges of travel cost. This can be expressed as follows :

$$AC^k_{i(r)} = \Sigma D^g_{j(r)} \tag{15}$$

subject to:

$$c^k_{ij} \geq \bar{c}^k \tag{16}$$

where: $AC^i_{i(r)}$ = accessibility of socio-economic group k in zone i to all urban opportunities of resource type r

$D^k_{i(r)}$ = number of opportunities of resource type r available to socio-economic group k in zone j

C^k_{ij} = cost of travel between i and j for socio-economic group k

$\bar{C}^k$ = mean regional generalised cost of travel for socio-economic group k.

The next measure is related to social deprivation. This measure involves a two stage sub-model based on a multiple linear regression relationship which predicts zonal median household income (Voorhees & Associates, 1968) and an exponential lognormal distribution of family income by zone (Goldner, 1972). The multiple regression relationship uses a lagged relationship as follows:

$$M_e X_{j_{t1}} = f(M_e X_{j_{t0}}, XR_{j_{t0}}, XC_{j_{(t0-t1)}} \dots) \tag{17}$$

where: $M_e X_{j_{t1}}$ = median household income of zone j at time t1

$M_e X_{j_{t0}}$ = lagged household income of zone j at time t0

$XR_{j_{t0}}$ = lagged net residential density of zone j at time t0

$XC_{j_{(t0-t1)}}$ = population change in zone j over period.

Using the median value as the position parameter for the log-normal function, the standard deviation of the lognormal distribution is estimated as follows:

$$\delta^2_j = [\Sigma_k \frac{(x^k_j - \bar{x}_j)^2}{n-1}] \tag{18}$$

where: δ_j = estimated variance of household incomes in zone j

$x_j k$ = income of household type k in zone j

n = number of households in zone j.

The zonal medians and standard deviations having been established for the zonal distribution of household income, the lognormal function can be expressed in terms of :

$$f(X_j) = (X_j \delta_j \sqrt{2\pi})^{-1} \exp \left\{ \frac{-(\ln x_j - \mu_j)}{2\delta_j^2} \right\} \qquad (19)$$

where : $f(X_j)$ = lognormal frequency distribution of an x income household in zone j

μ_j = natural logarithm of median household income in zone j

π = 3.1417

ln = natural logarithm with base of e.

The numbers of household in any one income band and zone can then be estimated by integrating (the frequency distribution) over the appropriate income band. This provides a useful measure by which to examine changes in both the concentration and spatial location of low income households across a range of different metropolitan policies.

V. Model Application

I will not digress on the model's empirical verification as this is well documented in the literature (Turner 1972; Federal Highways Administration, 1975). I have instead selected from various studies those aspects to the model's application which would seem to be of most relevance to this Conference. In so doing, it is useful to refer to figure 1 which summarises the chronological development of the USM family of models.

(1) Sub-regional Activity Allocation Model

This model was developed for the City of Bristol in the late 1960's to test alternative sub-regional growth strategies in responce to different sets of environmental constraints on development, such as liability to flooding, suitability of topography, aesthetic and recreational value of the landscape (Turner, 1970). The model was relatively crude in that the distribution of primary employment was generated exogenously, and it incorporated no efficiency or equity performance measures. Nevertheless, it was found to be of value both in assessing the probability of a given environmental policy being achieved in terms of the distribution of future growth relative to the specified environmental constraints and also of the energy "cost" of achieving such a policy. In terms of the energy "cost" of achieving a pattern of development consistent with the environmental constraints, the model predicted an increase in the mean journey-to-work travel time of some 7.75 percent, and in the service sector of 3.4 percent.

(2) Urban Systems Model/Urban Growth Simulation Model

This model was first applied in North America to the North Central Texas region as the Urban Systems Model. There it was used to assess the consequences of different public and private transportation policies on the region, the objective being to select and adopt a preferred multimodal regional transportation plan (North Central Texas Regional Policy Advisory Committee, 1974). For this purpose the primary employment distribution sub-model was developed and applied, as were measures of policy consistency, public utility investment, air pollution, energy consumption and resource accessibility.

(3) Statewide Activity Allocation Model.

This model represents a version of the USM developed specifically for the Federal Highway Administration for use in Statewide landuse and transportation policy analysis (Voorhees & Associates, 1974). The model was successfully calibrated for the State of Connecticut, and its sensitivity to different Statewide transportation policy assumptions examined. Highway system speeds were increased by up to 100 percent of those of the base year and decreased by 33 percent to reflect a wide range of different energy and investment policies. Constraints on development were expressed in terms of a recent designation of environmentally sensitive parts of the State. Policy impact was measured in terms of changes in the intensity of development, and the results indicated that a reduction in highway speeds of 33 percent would bring about a 7 percent rise in urban area population densities and a 10 percent fall in employment densities.

Speed increases of 100 percent caused a 15 percent fall in urban area population densities and a 15 percent rise in urban area employment densities. These variations were attributed to the effect of changes in residential sector holding capacities and competing population and employment centres.

VI. Future Model Development

For the sake of brevity I will confine my concluding remarks to the work being carried out in Texas, where the model has been successfully

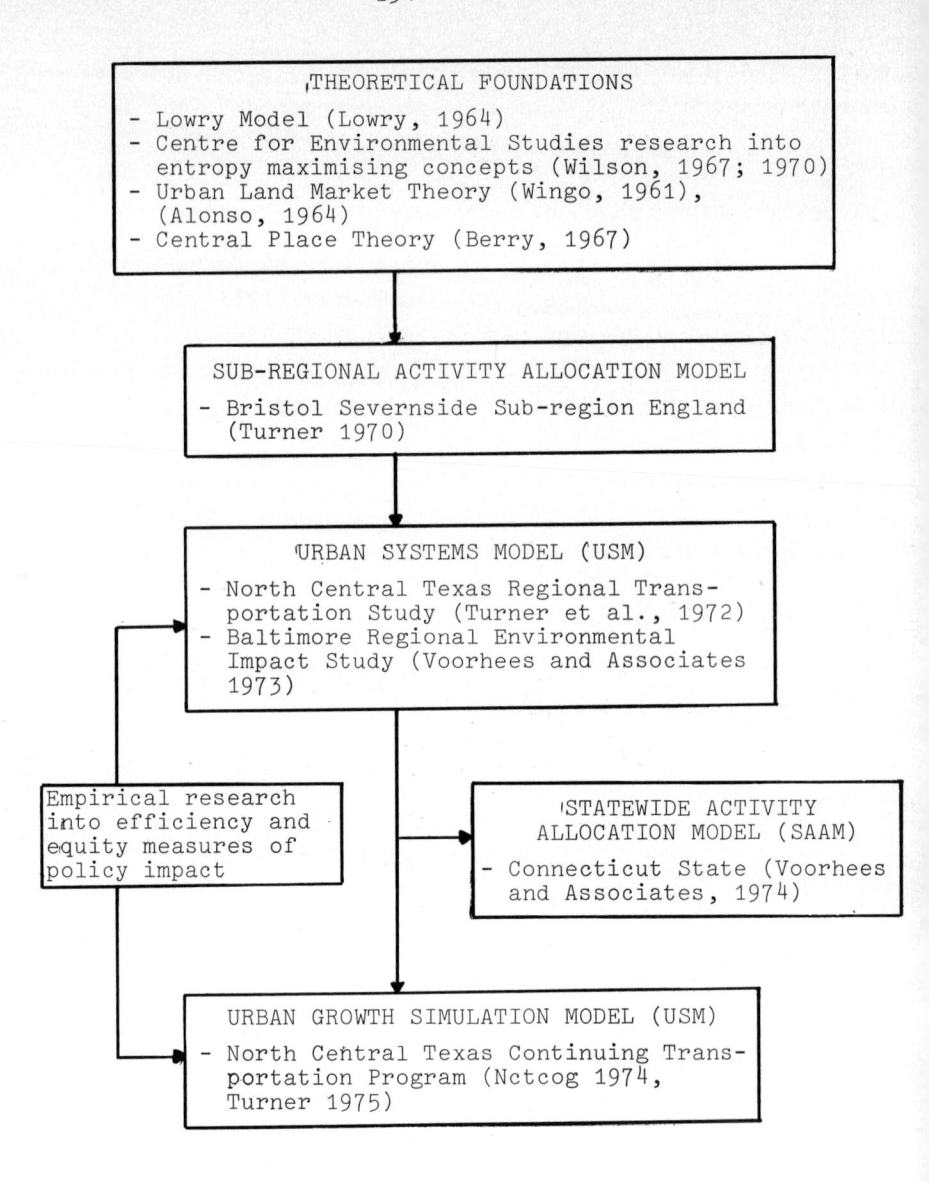

```
                    THEORETICAL FOUNDATIONS
        - Lowry Model (Lowry, 1964)
        - Centre for Environmental Studies research into
          entropy maximising concepts (Wilson, 1967; 1970)
        - Urban Land Market Theory (Wingo, 1961),
          (Alonso, 1964)
        - Central Place Theory (Berry, 1967)
                              │
                              ▼
             SUB-REGIONAL ACTIVITY ALLOCATION MODEL
           - Bristol Severnside Sub-region England
             (Turner 1970)
                              │
                              ▼
                    URBAN SYSTEMS MODEL (USM)
           - North Central Texas Regional Trans-
             portation Study (Turner et al., 1972)
           - Baltimore Regional Environmental
             Impact Study (Voorhees and Associates
             1973)

  ┌──────────────┐                    STATEWIDE ACTIVITY
  │Empirical     │                   ALLOCATION MODEL (SAAM)
  │research into │──►     - Connecticut State (Voorhees
  │efficiency and│           and Associates, 1974)
  │equity measures│
  │of policy     │
  │impact        │
  └──────────────┘
                    URBAN GROWTH SIMULATION MODEL (USM)
           - North Central Texas Continuing Trans-
             portation Program (Nctcog 1974,
             Turner 1975)
```

Figure 1 - Evolution of the Urban Systems Model Family

Source : Nathaniel Lichfield and Partners (1975).

applied to the regional transportation planning process and its predictions and impact measures have been used in the selection of a multi-modal regional transportation plan.

Indeed, it has been found that once the most suitable measures of the efficiency impacts of metropolitan transportation policy alternatives have been established, models such as the USM could probably provide some of this information far more cheaply and little less crudely than conventional transportation model procedures. In view of the broadening requirements of the urban transportation planning process this is significant. What is more, such models can provide a substantial amount of secondary information on the equity effects of policy alternatives that can be of use in their evaluation.

Work is now in hand to examine the robustness of the North Central Texas plan against a wide range of different assumptions. These include : the impact of a continued national economic recession and further increases in the cost and scarcity of energy resources; of alternative energy convervation policies; and of alternative environmental conservation policies on the future wellbeing of the regional community.

Each of these scenarios would be input to the model in the manner described through different transportation service levels, public utility investment or development constraint assumptions. The consequences would be traced through the model's predictions of future conditions and measurement of impact, to produce a robust and responsive regional transportation plan.

Footnote

1. It is further assumed that the perceived value of locational attractiveness lags behind its real value at a given point in time. This lag in the response of demand to the changing attractiveness of primary, residential and service sector locations is in part a function of imperfect information and communication, habit and inertia (Meier, 1962). For those reasons, the zonal attraction indices used in each of the activity system-sub-models are lagged behind in their real values for existing and future points in time (Cordey Hayes et al., 1970).

References

Alonso, W., 1964, Location and Land Use (Harvard Univeristy Press, Cambridge, Massachusetts).

Bellomo, S.J., 1972, Emergency and Growth of an Urban Region - the Developing Urban Detroit Area, Vol. 2: Future Alternatives, Detroit Edison Company.

Berry, B.J., 1967, Geography of Market Centers and Retail Distribution, (Prentice-Hall Inc., Englewood Cliffs, New Jersey).

Central unit for Environmental Planning, 1971, for Department of the Environmental), Severnside - A Feasibility Study, H.M.S.O..

Cordrey Hayes, 1970, M. and Massey, D.B., An Operational Urban Development Model of Cheshire, Working Paper No. 64, Centre for Environmental Studies, London.

Echenique, M., 1968, Urban Systems: Towards an Explorative Model, Workin Paper No. 7, Land Use and Built Form Studies, University of Cambridge, Cambridge.

Federal Highways Administration (prepared by Alan M. Voorhees ans Associates), 1974, Statewide Travel Forecasting Procedures including Activity Allocation and Weekend Travel, Phase II: Statewide Activity Allocation Model, Final Report.

Goldner, W., 1971, "The Spatial Distribution of Household Incomes", Paper presented at Tenth Annual Meeting of Western Regional Science Association, Berkeley.

Koppelman, F.S., 1969, A Model for Highway Needs Evaluation, ITR 4157-2490, Tri-State Transportation Commission.

Lichfield, N., Kettle, P. and Whitbread, M., 1975, Evaluation in the Planning Process (The Pergamon Press, Oxford).

Lowry, I.S., 1964, A Model of Metropolis, RM-4035-RC, The Rand Corporation, Santa Monica.

Meier, R., 1962, A Communications Theory of Urban Growth (M.I.T. Press, Cambridge, Massachusetts).

Neuburger, H., 1971, "User Benefit: Transport and Land Use Plans", Journal of Transport Economics and Policy, V., no. 1.

North Central Texas Steering Committee, Regional Transportation Advisory Committee, 1974, The Total Transportation Plan for the North Central Texas Region for 1990, Arlington, Texas.

Tiebout, C.M., 1962, The Community Economic Base Study, Paper No. 16, Committee for Economic Development, New York.

Turner, C.G., 1970, The Development of an Activity Allocation Model for the Bristol Subregion, Working Paper No. 8, Urban Systems Research Unit, Department of Geography, University of Reading, Reading.

Turner, C.G. et al., 1972, Application of the Urban Systems Model to a Region - North Central Texas, Alan M. Voorhees and Associates.

Turner, C.G., 1972, "A Model Framework for Transportation and Community Plan Analysis", Journal of the American Institute of Planners, 38.

Turner, C.G., 1975, "The Design of Urban Growth Models for Strategic Land-Use Transportation Studies", Regional Studies.

United States Department of Transportation, 1975, An Introduction to Urban Development Models and Guidelines for Use in Urban Transportation Planning, U.S. Federal Highways Administration.

Voorhees, Alan M. and Associates, 1968, Factors and Trends in Trip Lengths, N.C.H.R.P. Report No. 48, Highway Research Board.

Voorhees, Alan M. and Associates, 1973, Baltimore Regional Environmental Impact Study, Technical Memorandum Socio-Economic and Land Use Data.

Whitbread, M., Evaluation in the Planning Process - The Case of Accessibility, 1972, Working Paper No. 10, Planning Methodology Research Unit, School for Environmental Studies, University College London.

Wilson, A.G., 1967, A Statistical Theory of Spatial Distribution Models, Transportation Research, Volume 1.

Wilson, A.G., 1970, Entropy in Urban and Regional Modelling (Pion Ltd., London).

Wingo, L., 1961, Transportation and Urban Land, Resources for the Future Inc., Washington D.C.

MOTIVATION IN SUBURBAN MIGRATIONS RELATED TO ENVIRONMENTAL STANDARDS.
AN ANALYSIS OF THE ANTWERP REGIONAL MIGRATIONS.

M. Van Naelten
K.U.Leuven
Louvain - Belgium

I. A general framework for the analysis

In 1973 the Department of Town and Country Planning (Ministry of Public Works) ordered a study on long term strategy for the suburban development of the Antwerp Region. The ministry was especially concerned with the vanishing north-eastern greenbelt around the town and asked for the formulation of a general land-use policy[1].

The 3-year research program included three main topics in its report :

a) land-use and price policies;
b) environmental problems;
c) long term development strategy.

Suburban migration interferes with each of the topics. A tremendous difficulty in this matter was the total lack of recent data. The last census dated from 1961 and presents a very incomplete picture of migration movements.

It was obvious that a large-scale enquiry would be the only valuable alternative to bridging this information gap. The dimension of the enquiry was determined by several constraints such as the fact that the length of the intervieuw should not exceed a limit of one and a half hour, while the broad range of objectives required a great number of questions.

The publication of a research report of the American Highway Research Board (1969) at that time simplified the preparation of the enquiry.

An analysis of the enquiry, published in that report, convinced the advisors board of the study that the most operational approach was to use a translated and slightly adapted version of the enquiry list.

II. Some theoretical background

The study of intra-urban migration covers a large range of publications. Most urban models contain one or more blocks dealing with this continuously changing and renovating phenomenon. The meaning and the testing of an hypothesis certainly will differ according to the type of the model behind it. In spite of the subjectivity embedded in classifying urban models, one could tentatively present them as follows :

a) Gravitational and/or entropy models in which distance-minimizing options with or without constraints dominate the explanation of migration in the urban and suburban areas.
One could make a subclassification in these types of models in which land-use economic options (Alonso, 1964, Muth, 1969) play an importand role :

   i - market equilibrium models, in which location of economic activities and housing tend toward a dynamic general equilibrium. In these models consumers try to optimise their location by minimizing total costs, whatever these costs are (rents, transport costs, etc.). Examples are the Chapin-Weiss-Donally models (Chapin and S. Weiss, 1965), and the model of Metropolis (Lowry, 1964). The entropy models of A. Wilson (Wilson, 1970) generalise to some extent the Lowry models.

   ii - partial equilibrium models in which transportation problems dominate the development of the model. The Multiple Equation Model of J. Kain (Kain, 1962) emphasis the optimal commuting patterns and illustrates this type of models.

   iii - market demand models in which consumer behavior is evaluated in order to estimate size and quality of housing demand.

b) Feedback oriented behavioral models : in these models more interest is given to continuous adaptations and processes under changing conditions of this development is offered by the application of Forrester's approach to the Susquehanna River Planning and the famous "Club of Rome" reports.

Although feedback mechanisms can be detected in several rather more complicated equilibrium models (especially in some simulation models), the equations used normally do not contain adaptable thresholds, except in programming models of the Herbert-Stevens type (Herbert and Stevens, 1960).

An especially important aspect of these models is the impact of norms and perception rates on the subsequent behavior of the model over time.

This classification neither refers to the methods and techniques used, nor to the purposes of the applied models (description, prediction, etc.). It only provides a general background to our study.

III. Methods and enquiry contents

After deciding to use a sample, two elements became particularly important in our enquiry :

a) About 1.200 households that migrated into another commune during the last 2 years were interviewed;
b) The sampling design was adapted in order to test some aspects of the urban-rural gradient theory. In a previous study on the degrees of urbanity (Van Naelten, 1973) at the level of Belgian municipalities, six factorscores have been calculated from 48 variables. Using data for the Antwerp region, a cluster analysis was applied to make typological spatial regroupments. The method we used is the Rubin and Friedman (1967) Cluster analysis on covariance basis. The six clusters of communes have been regrouped into three circular areas around the central core of Antwerp :

- an urban area;
- a suburban area
- a rural but suburbanizing area.

Map 1 shows the distinct areas for which similar random sampling conditions were accepted.

From the enquiry we obtained information on the following range of topics :

a) Identification and location of the household, size, educational level of each member of the family;
b) Working and commuting conditions of each family member, costs, travel time, transportation mode;
c) Actual housing situations of the household, -ownership, housing costs, -intentions to build or to rent, evaluation of house and neighborhood, frequency of contacts, satisfaction and aspirations, physical characteristics of the house and the neighborhood, frequency of shopping at several hierarchical levels, etc.
d) Future housing situation projections : normative approach to future house and neighborhood conditions, attitudes towards compromises;
e) Evaluation of former housing conditions. (Similar information as in c);
f) Some additional information : recreation behavior, income levels, attitudes towards income, careerism, etc.. Finally, some psychosomatic observations.

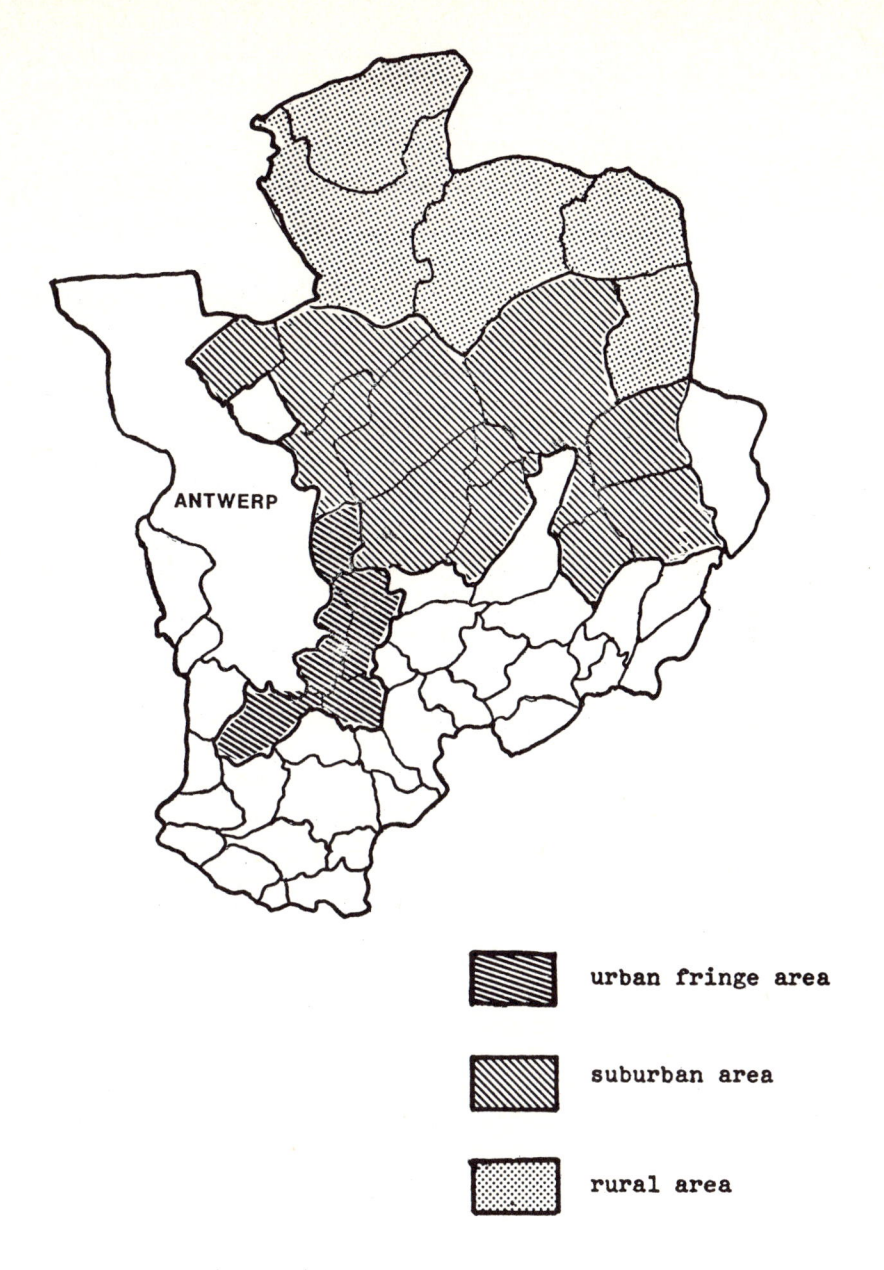

MAP. 1. : SAMPLE AREA

- urban fringe area
- suburban area
- rural area

The interviews provided a set of 585 variables on each household.

One possible approach might have been a simple testing of an hypothesis described in the available theoretical models. The data could have been analysed along the lines given by the Highway Research Report.
Nevertheless some doubts remained on the completeness of the main determining structures in migration analysis. Objectives and evaluations probably would interfere with the migration process, as will be illustrated in the final results of the analysis.

Nevertheless, there was no valid reason to believe that these objectives and evaluations would be especially homogeneous, as has been suggested in nearly all the existing models.

The lack of a clear and satisfactory theory on this point led to a less axiomatic approach, in which factor analysis was the main technique for detecting hidden structures and for generating new assumptions.
The total set of 585 variables was divided into 5 subsets :

    I variables : identification data, including data on commuting
    A variables : social attitudes and psychosomatic data
    M variables : migration data
    H variables : housing data
    B variables : neighborhood data

From these subsets four submatrices O_i (i=1,2,3,4) have been constructed.

a) Submatrix O_1 : I variables and A variables were grouped; from this set, by means of a first factor analysis, we constructed a simplified I-A subset. This set was joined to the remaining data to be analysed, and contained a selection of data with high communalities and high varimax loadings in one or two columns.

b) Submatrix O_2 : the simplified I-A set joined to the M variables.

c) Submatrix O_3 : the simplified I-A set joined to the H variables.

d) Submatrix O_4 : the simplified I-A set joined to the B variables.

Each submatrix contained about 170 to 190 variables. The main objective of this preliminary factor analysis was to detect the most characteristic variables, again on the basis of their communality and the level of varimax-loading values in one or more columns.

These variables were put together into a final matrix composed of :

    20 I variables   (Identification)
    10 A variables   (Attitudes)
    29 W variables   (Housing)
    47 B variables   (Neighborhood)
    32 M variables   (Migration)

The variance "explained" by the common factors is relatively low. Only 28 percent is explained by the 10 first factors. This result is not surprising. The analysis starts from a correlation matrix with different types of correlation coefficients : φ coefficients, point biserial coefficients, Pearson correlations.

This variety is caused by the mixing of dichotomous data, ranks and purely numerical values. Calculated on about 1.200 subjects, the correlation coefficients are low. Another point is the result of the effects caused by non-responses in the data sets.

One can doubt the value of factor-analysing these mixed types of data. Much depends on the way the results are finally used. Our main concern was only to detect unexpected structures or mechanisms and to test these structures by means of more reliable and "classic" methods. In the final matrix, only seven factors could be interpreted clearly :

a) one dominant factor shows the urban-rural gradient in which the suburban area characteristics have weaker loadings.
b) five factors each accentuating one main migration motive, each motive is associated with sets of several particular characteristics and attitudes.
c) a seventh factor is related to the general satisfaction and psychomatic characteristics; here a higher loading is found to be associated with the suburban area.

This formulation of migration motives resulted from open questions, so that these variables reflect a clear conscious response from the migrants. The important point was that the loading pattern in the varimax matrix suggested clearly a further deepening in the analysis of the data on the motive framework. This work has been done by simple cross-tabulations in which we calculated chi-square-tests for testing the hypothesis that significant differences existed between migrant groups, on the basis of their motives towards migration.

IV. Migration motives as a spatially differentiating factor

The main motives for migration mentioned in our sample are :

1. Moved for better housing   34,3 percent
2. Moved nearer to the job   22,3 percent
3. Just married   21,9 percent
4. Built or bought a house   14,7 percent
5. Obliged to move   6,8 percent

As shown in table 1 the spatial distribution of migrants according to their motive is different.

Table 1 : Migration motives per type of area, classified by increasing distance to the urban core

Motives	Urban area	Suburban area	Rural area
Better housing	26	34	26
Moved toward job	17	20	19
Just married	23	10	24
Built or bought a house	10	15	13
Obliged to move	7	5	4
Other reasons	17	16	14

Further analysis leads to the conclusion that :

- each type of migrant shows specific household characteristics, housing and working conditions, income classes and attitudes;
- each type has a particular living style and housing satisfaction pattern;
- each type of migrant uses differentiating needs, norms and evaluations toward actual and future housing conditions.

A complete description of the 54 crosstables should burden this paper. So our paper will only describe a general framework found in the data, on the basis of which we propose a taxonomy of intra-urban migration.

Intra-urban migrations can be subdivided into two main groups :

a) _free migrations_ : households have built or bought a house especially looking for better housing conditions. The households in this group on the average have more children, higher incomes and are highly sensitive to environmental qualities. These migrations mostly are final migrations toward an area between 10 to 15 km from the urban core. Ownership is a predominant characteristic.

b) <u>forced migrations</u> : one can distinghuish two subclasses in this group :
   - the first one contains migrants <u>forced to move by their landlords</u> and <u>newly married couples</u> looking for a house. In this subclass the impact of the neighborhood, relationships, nearness of family and rent level are predominant evaluation criteria.
   - the second subclass regroups households that <u>moved nearer to their jobs</u>; mostly, long distance migrations are found in this category; it is striking that price constraints and evaluation norms are mostly weaker in this subclass. The need for a shorter commuting trip seems to "sublime" other possible disadvantages of housing and neighborhood conditions.

A general characteristic of the forced migrants group is their preference for urban or rural locations and the provisional character of their actual location.

In spite of the rather large size of the sample, an important lack of information on the behavior of an eventual reference group of non migrants remains. As mentioned, only recent migrants from areas around Antwerp have been put in the sample. Nevertheless, our observations allow for a final conclusion on the particular role of the motivation to migrate.

V. <u>Conclusions and general remarks</u>

Although urban modelling remains an important tool in urban planning, some obvious oversimplifications often treaten this type of approach. As Harris points out (Harris, 1972) "density, housing, quality and neighborhood characteristics do not enter in a large number of urban models". But even if these variables were taken into account, some weaknesses would remain. It is not only a matter of introducing these variables in the analysis but also of perception by the population concerned especially when they are looking for a residential relocation. Furthermore, as can be derived from our analysis, groups of migrants perceive their actual, and especially their projected housing conditions in a totally different way, depending on their motives for moving to one or another location.

Of course, the introduction of these considerations into our actual models that pretend to simulate urban developments, will certainly complicate these already sophisticated approaches.

Nevertheless, the important differentiations shown in this enquiry, questions the value of predictions that are mainly based on assumptions of minimizing commuting and other distances. It is obvious that these objectives play an important role. Nevertheless, a refining of the existing models is necessary for a realistic forecasting of urban development planning.

Footnotes

1. This paper is based on work done under contract of the Ministry of Public Works of Belgium. I would like to thank the advisors board in this part of the study :
De Pessemier, E. : architect, D'Olieslager, L. : Dr. of Pol. and Soc. Science, Heremans, R. : Lic. Soc. Science, Lorent, J. : Lic. Geog. Science - Ministry of Public Works, Maes, M. : Lic. Math. Science - systems analist, Rosseel, E. : Drs. Psych. Science, Somers, H. : Dr. Psych. Science, Stevens, M. : Lic. Geog. Science - Ministry of Public Works, Van Acker, M. : Drs. Psych. Science, Smets, G. : Dr. Psych. Science.
However, I fully accept the responsibility of errors and omissions.

References

Alonso, W., 1970, <u>Location and Land Use : Toward a General Theory of Land Rent</u> (Harvard University Press, Cambridge, Massachusetts).

Butler, E.W., a.o., 1969, "Moving Behaviour and Residential Choice : a national survey", Highway Research Board, Report 81.

Friedman, H. and J. Rubin, 1967, "A Cluster Analysis and Taxonomy System for Grouping and Classifying Data", I.B.M. Corporation, New York.

Donelly, T.G., Chapin, F.S., Jr., and S.F. Weiss, S.F., 1964, "A Probabilistic Model for Residential Growth", Institute for Research in Social Science, University of North Carolina.

Harris, B., 1972, "A Model of Household Location Preferences", in Recent <u>Developments in Regional Science</u>, ed. Funk, R., (Pion, London), pp. 63-79.

Herbert, J.P. and B.H. Stevens, 1960, A Model for the Distribution of Residential Activity in Urban Areas, <u>Journal of Reg. Science</u>, II, no 2, pp. 21-36.

Kain, J.F., 1961, "The Journey to Work as a Determinant of Residential Location", The Rand Corporation, Santa Monica, California.

Lowry, I.S., 1964, "Model of Metropolis", The Rand Corporation, Santa Monica, California.

Muth, R.F., 1969, Cities and Housing : The Spatial Pattern of Urban Residential Land Use (The University of Chicago Press, Chicago and London).

Van Naelten, M., 1973, "Differentiële woonappreciatie naar type en motivering van de migraties", Part 3.2., in Planologische begeleiding van de suburbanisatie in de Voorkempen, Ministry of Public Works, Brussels.

Wilson, A.G., 1970, "Generalizing the Lowry Model", Working Paper 56, Center for Environmental Studies, London.

ON MULTI-REGIONAL MODELLING

J.H.P. Paelinck with A. Van Delft,
L. Hordijk and A.P. Mastenbroek
Netherlands Economic Institute
Rotterdam - The Netherlands

I. Introduction

Spatial economic relations cannot be composed in such a straightforward manner as would appear possible at first sight.

In most of the literature, one encounters regional and urban models that hardly integrate any spatial factor at all. At best, the level or evolution of some variables is explained by the level or evolution of other variables measured in the same area. The customary way to regionalize the exercises is by putting in spatial interaction in the form of distance frictions; this leads up to the concepts of potential, gravity, and input-output access. Areas seldom come into play, however, the relations often being of the point-point type still.

In estimating spatial economic relations, classical cross-sectional analysis is usually applied. In previous papers which introduced spatial connections (Hordijk and Paelinck, 1974 and 1975), it was found that the resulting econometric problem has specific features, and requires the development of adequate estimators for the parameters.

In this paper a number of considerations will be devoted to some formal representations in space, linking them up with a series of previously developed models. The only aim of the paper is to get an insight into the way spatial (especially multi-regional, and more specifically interregional) models should be set up rigorously, i.a. in connection with ecological problems.

Some contributions have been made to better modelling of spatial economic relations. Isard and Liossatos (1975) have looked for some parallels with space-time relations in physics. Like Beckman (1970, 1971), they have warned against a purely analogous interpretation of the relations uncovered by physicists. If, e.g., wave relations (Meriam, 1971) are observed, the economist should try to obtain a wave equation as the result of economic spatial interaction; deformation and translation in time-space, more generally, should have the same explanatory origin (Paelinck, 1973).

Section two of this paper describes a wave-like consumption pattern obtained from a simple discontinuous, static consumption model. In section three, a continuous differential system is investigated, to describe a wave-like motion of a given variable through time and space. The links between existing spatial economic models and the continuous relationships in space-time developed in section three, are explored in the following section. In section five some examples are worked out.

II. Spatial economic relations : a static example

Take the following simple static model :

$$y_r = c_r + j_r \qquad (II.1)$$

$$c_r = \alpha y_r + \beta \Sigma_\rho y_\rho + \gamma \Sigma_{\rho^*} y_{\rho^*} + \delta + \varepsilon_r \qquad (II.2)$$

where y_r is income, c_r consumption and j_r investments, all in region r; ρ is the index for regions contiguous of order 1, ρ^* the index for regions contiguous of order 2; ε_r is a stochastic term.

When the regions are situated as follows (figure 1)

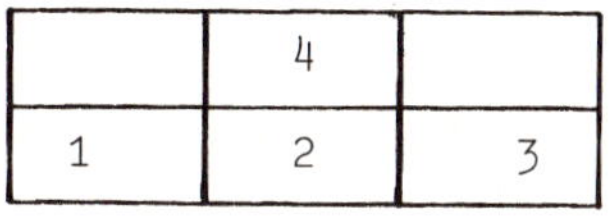

	4	
1	2	3

Figure 1

(II.1) and (II.2) can be rewritten in matrix notation as :

$$\underline{y} = \alpha \begin{bmatrix} 1 & 0 & 0 & 0 \\ 0 & 1 & 0 & 0 \\ 0 & 0 & 1 & 0 \\ 0 & 0 & 0 & 1 \end{bmatrix} \underline{y} + \beta \begin{bmatrix} 0 & 1 & 0 & 1 \\ 1 & 0 & 1 & 1 \\ 0 & 1 & 0 & 1 \\ 1 & 1 & 1 & 0 \end{bmatrix} \underline{y} + \gamma \begin{bmatrix} 0 & 0 & 1 & 0 \\ 0 & 0 & 0 & 0 \\ 1 & 0 & 0 & 0 \\ 0 & 0 & 0 & 0 \end{bmatrix} \underline{y} + \underline{j} + \delta \underline{i} + \underline{\varepsilon} \qquad (II.3)$$

where $\underline{y}$ represents a column vector with elements y_r; $\underline{i}$ is the unit column vector.

The matrices following β and γ are the first- and second-order contiguity matrices, respectively (with zeros on the main diagonal), henceforth to be symbolized by C_1 and C_2.

For the more general case, (II.3) becomes :

$$\underline{y} = \alpha I \underline{y} + \beta C_1 \underline{y} + \gamma C_2 \underline{y} + \underline{j} + \delta \underline{i} + \underline{\varepsilon} \qquad (II.4)$$

In order to guarantee consistency in the system, some small adaptations (Hordijk and Paelinck, 1974) are necessary. After these transformations (II.4) becomes :

$$\underline{y} = \alpha I \underline{y} + \beta^* C_1 \underline{y} + \gamma^* C_2 \underline{y} + \underline{j} + \delta \underline{i} + \underline{\varepsilon} \qquad (II.5)$$

The reduced form of this equation can be written as :

$$(I - \alpha I - \beta^* C_1 - \gamma^* C_2)\underline{y} = \underline{j} + \delta \underline{i} + \underline{\varepsilon} \qquad (II.6)$$

or

$$\underline{y} = M^{-1}(\underline{j} + \delta \underline{i} + \varepsilon) \qquad (II.7)$$

where $M = (I - \alpha I - \beta^* C_1 - \gamma^* C_2)$ \hfill (II.8)

is the spatial multiplier of the system. Its existence is due, in the simple model considered, to the fact that people have a tendency to spend their incomes in other regions than their own, in fractions declining with increasing distance.

That this consumption pattern can lead to wave-like results on income in space can be illustrated with the help of the following example.

Suppose the regions are located in space as shown in figure 2 below

1	2	3

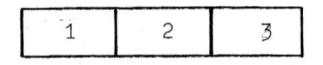

Figure 2

In that case M^{-1} looks like :

$$M^{-1} = \begin{bmatrix} 1 & -a & -b \\ -a & 1 & -a \\ -b & -a & 1 \end{bmatrix}^{-1}$$

For values of $a = 0.6$ and $b = 0.2$, one can compute M^{-1} as

$$M^{-1} = \frac{1}{0.096} \begin{bmatrix} 0.64 & 0.72 & 0.56 \\ 0.72 & 0.96 & 0.72 \\ 0.56 & 0.72 & 0.64 \end{bmatrix}$$

These values of m_{ij} give the total (i.e. direct and indirect) result in region i of the system, of a unit impulse exerted on region j. As a partial consequence of the assumption that the fraction spent in regions of first-order contiguity is higher than the fraction spent in second-order contiguous regions, the total effect on region 2 is even stronger than that in region 1, where an initial impulse is supposed to have been given. Generalizing, it does not seem unreasonable to suppose that the total effect on a region is higher as the number of first-order contiguity regions is higher. For that reason, core regions have some advantages over peripherical regions; however, as the total income has to be spent somewhere, it can hardly be expected that parameters α, β and γ are alike for peripherical and core regions; this essential asymmetry in real geographical space will not be pursued here, however.

Generalization to a set of regions is immediate; column vectors of matrix M^{-1} ($\underline{m}_j$) depict the total transmission of a unit impulse in a region (j), row vectors ($\underline{m}'_i$) the effect on region i of unit impulses given elsewhere.

Study of matrices of type M (related to more complex models, of course, than the one set up here) would actually reveal unsuspected spatial transmission patterns.

The multiplicative power of M^{-1} depends on two factors :
- its coefficients;
- the interregional linkages, in the case of model (II.4) revealed by matrices C_1 and C_2.

A possible theoretical analogue would be a generalized band matrix (Bodewig, 1959) of type

$$M = I - B_1 - \Gamma_1 \qquad (II.9)$$

where B_1 would be an approximation to $\beta^* C_1$ and present itself as a corresponding number of upper and lower diagonal "bands" around the principal diagonal I, and Γ_1 would border the B_1-bands likewise.

A well-known case is the auto-correlation inverse variance matrix (Theil, 1976) in econometrics, where ρ refers to the parameter of a first-order autoregressive model,

$$V^{-1} = \begin{bmatrix} 1 & -\rho & 0 & \cdots & 0 \\ -\rho & 1+\rho^2 & -\rho & \cdots & 0 \\ 0 & -\rho & 1+\rho^2 & \cdots & 0 \\ \vdots & & & \ddots & \vdots \\ 0 & 0 & 0 & \cdots & 1 \end{bmatrix} \qquad (II.10)$$

which is the inverse form

$$V = (1-\rho^2)^{-1} \begin{bmatrix} 1 & \rho & \rho^2 & \cdots \\ \rho & 1 & \rho & \cdots \\ \rho^2 & \rho & 1 & \cdots \\ \vdots & \vdots & \vdots & \\ & & & 1 \end{bmatrix} \qquad (II.11)$$

Given that $\rho < 1$, and identifying V^{-1} with M, the resulting multiplier M^{-1} would give a series of decreasing effects in space. As is evident from (II.10) and (II.11), the multiplier M^{-1} is spatially fully connected, which is due to the fact that V^{-1} is indecomposable.

As band matrices are symmetric, their characteristic values are real, which means that, unlike in dynamic (temporal) models, no link can be traced between the complex character of eigenvalues and cyclical behaviour (Gandolfo, 1971).

To get an idea why generalized band matrices can generate cycles, consider the matrix with structure

$$\underline{a}' = ..0 \quad 0 \quad -b \quad -a \quad 1 \quad -a \quad -b \quad 0 \quad 0 \quad \cdots \qquad (II.12)$$

For any column $\underline{m}_i$ of M^{-1} there obtains

$$\underline{a}'_j \underline{m}_i = 0, \qquad j \neq i. \qquad (II.13)$$

Also

$$\underline{a}'_j \Delta \underline{m}_i = 0 \qquad (II.14)$$

and the same for higher-order differences.

This explains why (II.11) obtains, as a sufficient condition for (II.13) - (II.14) to be satisfied is

$$\underline{\Delta m}_i \propto \underline{m}_i \qquad (II.15)$$

which is the case for vector $[1, \rho, \rho^2, ...]$.

Taking three differences in expression (II.14) and supposing terms farther away to be negligible, one computes

$$\Delta_{23} = -a^{-1} b \Delta_{12} \qquad (II.16a)$$

$$\Delta_{34} = a^{-2} b \Delta_{12} \qquad (II.16b)$$

showing that Δ_{23} and Δ_{34} have, respectively, the opposite and the same sign as Δ_{12}.

In fact, exact generalized band matrices can only be obtained in cases where the spatial layout of regions follows figure 2 (linear structure); in more general cases (nested triangles, squares, hexagons) one obtains much more complex matrices M, in which neighbouring cells are not necessarily representative of neighbouring regions.

This raises the problem of a <u>more adequate representation</u> of the spatial structure. In fact, very often the geographical structure is immediately amenable to matrix representation, as is the case with a nested-squares structure (figure 3).

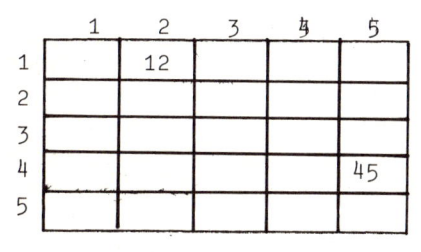

	1	2	3	4	5
1		12			
2					
3					
4					45
5					

Figure 3

A tentative algebra to try would be a Kronecker-Schur matrix algebra, of which only some outlines will be given here.

Define a Kronecker-Schur product as

$$[C_{ij}] = [B_{ij}] a_{ij} \qquad (II.17)$$

or

$$C = B \ (ks) \ A \qquad (II.18)$$

In our case (figure 3) A is square (5 x 5) and B and C are square (25 x 25); each B_{ij} square 5 x 5 matrix is scalarly multiplied by the corresponding a_{ij}-term of A.

Result (II.7) can then be expressed as

$$Y = M^{-1} \ (ks)E \qquad (II.19)$$

where E is the matrix of exogenous impulses in each ij-region; $Y_{ij} \epsilon Y$ a 5 x 5 matrix, gives the total direct plus indirect effect on all regions of an impulse e_{ij} given in region ij. In fact, $M_{ij}^{-1} \epsilon M^{-1}$ results from a matrix representation of the columns of M^{-1} (expression II.7) corresponding to region ij.

How (II.19) is to be derived directly from the initial expression,

$$[(1-\alpha)C_0 - \beta C_1 - \gamma C_2)] \ (s)Y_0 = C_0(ks)E \qquad (II.20)$$

has still to be investigated.

Another approach, however, is the more classical vector representation of Y and E, vec Y and vec E, where the successive columns of Y and E (n x n) are arranged in an (n x n) x 1 vector (see e.g. Fisk, 1967).

Scheme 4 that follows gives the condensed contiguity matrix (orders 0,1 and 2) for a 4 x 4 system of squares; black dots or circles represent absence of these relations; only th right upper triangle of the symmetric matrix is represented.

From Figure 4 one can easily compute that

$$M_{31} - M_{21} = -A_{11}^{-1} \ [(A_{12} - A_{13})(M_{32} - M_{22})] \qquad ^{1)} \qquad (II.21)$$

Here A_{11}^{-1} and $(A_{12} - A_{13})$ are non-negative, in the case of a Leontief-Minskowski type of structure; further diag M_{22} > diag M_{32}, and if this diagonal is dominating, $M_{31} - M_{21} \geq 0$, showing that total effects, farther away in space, can be higher than effects nearby, again a wave-like motion in space.

	1	2	3	4	5	6	7	8	9	10	11	12	13	14	15	16
1	0	1	2	.	1	1	2	.	2	2	2	.				
2		0	1	2	1	1	1	2	2	2	2	2			●	
3			0	1	2	1	1	1	2	2	2	2				
4				0	.	2	1	1	.	2	2	2				
5					0	1	2	.	1	1	2	.	2	2	2	.
6						0	1	2	1	1	1	2	2	2	2	2
7							0	1	2	1	1	1	2	2	2	2
8								0	.	2	1	1	.	2	2	2
9									0	1	2	.	1	1	2	.
10										0	1	2	1	1	1	2
11											0	1	2	1	1	1
12												0	.	2	1	1
13													0	1	2	.
14														0	1	2
15															0	1
16																0

Figure 4

A short proof for the structure of the scheme in figure 4 is offered here to conclude. Take the first row, for instance; if the original matrix was (n x n), the first n terms will show the appropriate contiguity relations for term (1,1) in increasing order, starting, of course, with 0; the next n terms will do so again, starting with 1, and so on, until the relevant contiguity effects have been exhausted. The second row will do alike for term (2,1), and so on.

It is conjectured that this vector development of spatial matrices might be a fruitful tool for further investigation; the matter will be treated as such in the future. The way in which empirically observed systems of regions can be formalized along the lines sketched will also be subject matter for research.

III. Economic time-space relations

Section II contained a <u>static</u> example of <u>discrete</u> spatial economic relations. It is the purpose of this section to switch to <u>dynamics</u> and to <u>continuous</u> space. A starting point for the investigation could be the following differential system, in one variable z [2]).

$$\dot{z} = f_1(x, y, t; \underline{u}) \qquad (III.1)$$

$$z_x = f_2(x, y, t; \underline{u}) \qquad (III.2)$$

$$z_y = f_3(x, y, t; \underline{u}) \qquad (III.3)$$

where $\dot{z} \triangleq \partial z/\partial t$, $z_x \triangleq \partial z/\partial x$ and $z_y \triangleq \partial z/\partial y$.

The system has been borrowed from optimal control theory (see e.g. Strauss, 1968) and generalized to partial derivatives with respect to the spatial coordinates x and y; $\underline{u}$ is a vector of control variables, to which we shall come back.

Fixing $\underline{u}$ for the time being, one can imagine system (III.1) through (III.3) to be integrated with respect to x, y and t and give a function

$$z = z(x, y, t) \quad (III.4)$$

describing the behaviour of the variable z in space and time.

A simple example is the following function :

$$z(x,y,t) = \gamma \sin \frac{\pi}{2} [\alpha(t-t_o) + \beta\sqrt{(x-x_o)^2+(y-y_o)^2}] \quad (III.5)$$

describing a wave-like motion of z through time and space. Equation (III.5) verifies indeed the fundamental wave equation (Meriam, 1971).

$$\partial^2 z/\partial t^2 = c^2 \partial^2 z/\partial r^2 \quad (III.6)$$

with $r = \sqrt{(x-x_o)^2 + (y-y_o)^2}$ and $c = \alpha\beta^{-1}$.

For a given fixed point (e.g. x_o, y_o), (III.5) describes the fluctuation of variable z through time (see figure 5).

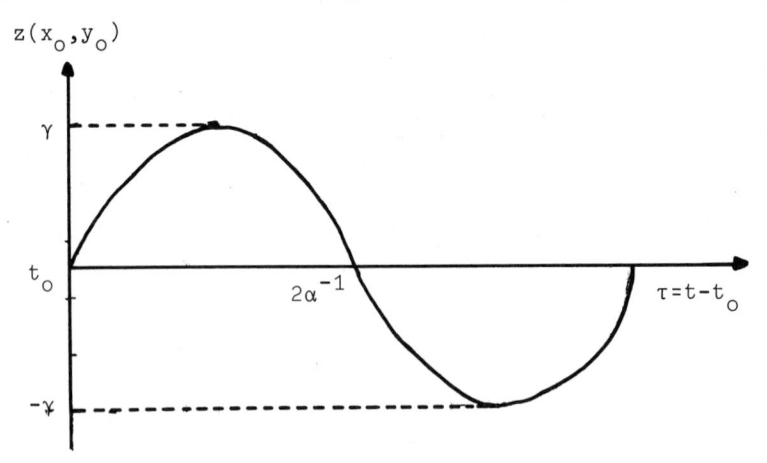

Figure 5

For given t, e.g. $t=t_o$, (III.5) also describes what happens in space over a circle with radius $r = \sqrt{(x-x_o)^2 + (y-y_o)^2}$ (see figure 6).

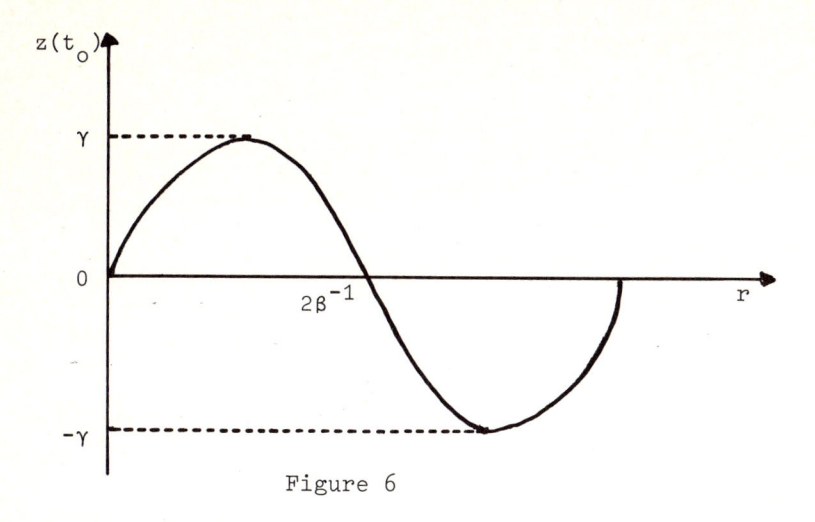

[Figure 6: Graph showing $z(t_o)$ versus r, with horizontal dashed lines at γ and $-\gamma$, and a marked point at $2\beta^{-1}$ on the r-axis. The curve rises to γ, crosses zero, dips to $-\gamma$, and returns toward zero.]

Figure 6

What is important, however, is to analyse under what economic conditions equation (III.6) originates; moreover, as policy measures intervene in it, it will be hard, in the absence of a simple time pattern $\underline{u}(t)$, to integrate system (III.1), (III.2), (III.3), so that equation (III.6) probably will not emerge.

In this sense equations (III.1), (III.2) and (III.3) are more useful, as they allow of writing the problem on hand in typical control terms, introducing a performance criterion :

$$\text{Max } J = \int_S \int_0^T d(x,y,t)\, f[\,z(x,y,t;\underline{u}),\underline{u}(x,y,t)\,]\,dxdydt \qquad (III.7)$$

where S is the relevant area, T a time horizon, $d(x,y,t)$ a general space-time discount function, and $f(z,\underline{u})$ an evaluation function of the state and control variables over space and time.

A problem creeps up, however : in a dynamic setting, is there any use for partial derivaties z_x and z_y ? One could probably confine oneself to function (III.1) and an initial distribution of z, z_o.

$$z_o = z_s \cdot g(x,y,t_o)\,dx\,dy \qquad (III.8)$$

where z_s is the sum total over S, and g the spatial density function at $t = t_o$, with

$$\int_S \int g(x,y,t_o)\,dxdy = 1 \qquad (III.9)$$

(III.1), (III.8) and (III.9) thus having been stated, derivatives z_x and z_y follow immediately and further specification, as in (III.2) and (III.3), is not necessary.

However, in some cases one can imagine z_x and z_y to appear even in a dynamic function. Take the case, for instance, of an environmental control measure u_i, preventing waste material from being drained via rivers to other regions, or smoke from being freely exhausted via chimneys. Given the presence of u_i, locations in the direct neighbourhood of the emission point (x,y), should be free from waste or should not exceed a certain proportion α^* of the level of pollution in point (x,y), $z(x,y)$, say. Translated into mathematical terms this means

$$z_x = z_y = -(1-\alpha^*)z(x,y) \qquad (III.10)$$

To show that (III.1), (III.2) and (III.3) can easily accommodate our classical regional models, a number of examples will be given now.

IV. Links with existing models

Let us first take up the static consumption model of section II. Equation (II.5) describes how income in a certain region depends on income in other (surrounding) regions and on investment in the region itself. Switching from discrete space to continuous space, (II.5) can be rewritten as :

$$z(x,y) = \int_S\int f[x,y,\xi,\eta\,;\ z(\xi,\eta)]\,d\xi d\eta + k(x,y) \qquad (IV.1)$$

Equation (IV.1) is applicable to all points (x,y) in space S. Symbol z means income, symbols ξ and η denote a point in S, and $k(x,y)$ is the continuous version of the last part of equation (II.5).

An example of f would be :

$$f = z(\xi,\eta)e^{-\gamma r}, \qquad (IV.2)$$

with $r = \sqrt{(\xi-x)^2 + (\eta-y)^2}$

Obviously, additivity problems would be present, as (IV.1) integrated over S would have to reproduce the total (national) value of $z^{3)}$.

Total exports of good i out of a region dxdy, $n_i(x,y)$, in the Lentief-Strout gravity (Leontief and Strout, 1966) model would be

$$n_i(x,y) = p_i(x,y) \int_S \int f[(x,y,\xi,\eta;d_i(\xi,\eta)] \, d\xi d\eta \qquad (IV.3)$$

where p would be production (push-factor), d demand (pull-factor) and f a friction function as in (IV.1)-(IV.2).

A similar formulation could be given of Klàassen's multi-regional static attraction model (Klaassen and van Wickeren, 1969), where

$$p_i(x,y) = \sum_j \int_S \int g_j[x,y,\xi,\eta; d_j(\xi,\eta)] \, d\xi d\eta \qquad (IV.4)$$

the g_j being attraction functions, and the indices i and j referring to goods and services. The dynamic Paelinck version (Paelinck, 1973) would read

$$\dot{p}_i(x,y) = \sum_j \int_S \int g_j^*[x,y,\xi,\eta;d_j(\xi,\dot{\eta})] \, d\xi d\eta \qquad (IV.5)$$

where $\dot{p}_i \triangleq \partial p_i / \partial t$.

Equations (IV.4) and (IV.5) are simplified, in the sense that forcing factors (policy measures) and other locational factors have been neglected.

As expression (IV.5) seems to be typical for recent location models, some developments will be presented as to how it could be used in theoretical research on interregional dynamics.

V. Uncovering spatial growth theorems

Dealing with continuous problems, let us investigate the time-space behaviour of a state variable z obeying

$$\dot{z}(x,\dot{y}) = a \int_S \int z(x,y) e^{-b\sqrt{(x-\xi)^2+(y-\eta)^2}} \, d\xi d\eta \qquad (V.1)$$

An example would be the integration over a circle, computing the growth of z in each point as a function of the potential measured in that point, something which is closely linked up with model (IV.5).

Let S be a circle with radius R and a uniform initial density for z, $z_0 = 1$; suppose further that[4]

$$\dot{z}_0(\rho) = \int_0^\Pi \int_0^{\rho^*(\phi)} r \, e^{-r^2} \, dr d\varphi = \int_0^\Pi \int_0^{\rho^*(\varphi)} r \, e^{-r^2} \, dr d\omega \qquad (V.2)$$

where $\rho^*(\varphi)$ can be computed as (see figure 7), by the cosine rule

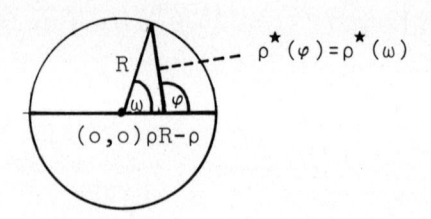

Figure 7

$$\rho^*(\varphi) = -\rho \cos \varphi + \sqrt{\rho^2(\cos^2\varphi - 1) + R^2} = \sqrt{R^2 + \rho^2 - 2\rho R \cos} = \rho^*(\omega) \quad (V.3)$$

Integration over r gives

$$\dot{z}_o(\rho) = -\frac{1}{2}\int_0^\Pi [e^{-r^2}]_0^{\rho^*(\omega)} d\omega \quad (V.4)$$

$$= \frac{\Pi}{2} - \frac{1}{2}\int_0^\Pi e^{-\rho^{*2}(\omega)} d\omega \quad (V.5)$$

We should like to compute $\frac{\partial \dot{z}_o(\rho)}{\partial \rho}$, suspecting that this expression is negative, so that central regions would grow faster than peripheral ones from the very start of the process on.

Given that (V.5) is a continuous function, one may interchange the derivative and integral operations. There results

$$\frac{\partial \dot{z}(\rho)}{\partial \rho} = \int_0^\Pi e^{-\rho^{*2}(\omega)} \rho^*(\omega) \frac{\partial \rho^*(\omega)}{\partial \rho} d\omega \quad (V.6)$$

$$= e^{-(R^2+1)} \int_0^\Pi e^{2R\cos\omega}(1 - R\cos\omega) d\omega \quad (V.7)$$

where ρ has been put equal to 1, implying $R > 1$. Expanding $e^{2R\cos\omega}$ in a McLaurin series, and dropping zero terms (Hodgman, 1961), one gets

$$e^{R^2+1} \cdot \frac{\partial \dot{z}(\rho)}{\partial \rho} = \int_0^\Pi d\omega + R^2(\frac{2^4}{4!} - \frac{2^3}{3!}) \int_0^\Pi \cos^4\omega \, d\omega$$

$$+ R^2(\frac{2^6}{6!} - \frac{2^5}{5!}) \int_0^\Pi \cos^6\omega \, d\omega + \ldots \quad (V.8)$$

$$= -\sum_{n=0}^\infty \frac{\Pi(R)^{2n}(n-1)}{(n!)^2} \quad (V.9)$$

In (V.9) the term for $n = 0$ is equal to Π, for $n = 1$ equal to zero;

the following terms, all negative, form a progression with first term $R^4/4$ and a ratio r declining with n

$$r = \frac{nR^{2n-4}}{(n+1)^2(n-1)} \qquad (V.10)$$

The sum (V.9), starting from $n = 2$, is bounded from above by the sum of the geometrical progression with ratio equal to (V.10) for $n = 2$, i.e. :

$$S = -\frac{R^4 \Pi}{4}[1 + 2/9\ R^2 + (2/9\ R^2)^2 + \ldots] \qquad (V.11)$$

$$= -\frac{9 \cdot R^4 \Pi}{36 - 9R^2} \qquad (V.12)$$

which expression can be made larger than $-\Pi$ for R near enough to 1.

We can conclude that $\partial \dot{z}(\rho)/\partial \rho < 0$ nearly everywhere over R, except for a fringe crown near the limit of the circle.

VI. Conclusions

This paper presents in embryonic form some ideas about the problem how to specify regional economic relations. The authors again want to emphasize the importance of this specification problem. Spatial interactions should appear systematically in the treatment of regional economic phenomena; the spatial income model in section II is an example of how these interrelations can be modelled. Further work on this kind of model has to be done, especially about presumed spatial behaviour of economic actors underlying the model.

The idea to use a vector-matrix-representation in which real geographical structure can be maintained is in its first stage of development. One has to develop more deeply the ideas mentioned in this paper, especially those which are connected with the Kronecker-Schur product calculus.

Footnotes

1. The scheme of figure 4 can be written as
$$A = \begin{bmatrix} A_{11} & A_{12} & A_{13} & A_{14} \\ \cdot & \cdot & \cdot & \cdot \\ \cdot & \cdot & \cdot & \cdot \\ A_{41} & \cdot & \cdot & A_{44} \end{bmatrix}$$
where $A_{11} = A_{22} = A_{33} = A_{44}$; $A_{12} = A_{23} = A_{34} = A_{43} = A_{32} = A_{21}$, and $A_{13} = A_{24} = A_{42} = A_{31}$. By multiplying
$$A = \begin{bmatrix} A_{11} & \cdots & A_{14} \\ \cdot & & \cdot \\ \cdot & & \cdot \\ A_{41} & \cdots & A_{44} \end{bmatrix} \text{ by } A^{-1} = M = \begin{bmatrix} M_{11} & \cdots & M_{14} \\ \cdot & & \cdot \\ \cdot & & \cdot \\ M_{41} & \cdots & M_{44} \end{bmatrix}, \text{ it is possible to derive (II.21)}.$$

2. Generalization to a vector $\underline{z}$ is straightforward.
3. For the discrete case, see Hordijk and Paelinck (1974).
4. The square root has been neglected, so that quadratic distances are considered. The expression as a function of ω has been suggested by Leo Klaassen and will be used for further derivations.

References

Beckman, M., 1970, "The Anlysis of Spatial Diffusion Processes", <u>Papers and Proceedings of the Regional Science Association</u>, Vol. 25, pp. 109-117.

Beckman, M., 1971, "Some Aspects of Economic Diffusion Processes", in : H.W. Kuhn and G.P. Szegö, <u>Games and Related Topics</u>, (North Holland Publishing Company, Amsterdam), pp. 313-323

Bellman, R., 1960, <u>Introduction to Matrix Calculus</u>, MacGraw Hill, New York.

Bodewig, E., 1959, <u>Matrix Calculus</u>, (North Holland Publishing Company, Amsterdam) (2nd revised edition).

Fisk, P.R., 1967, <u>Stochastically Dependent Equations</u> (Griffin, London).

Gandolfo, G., 1971, <u>Mathematical Methods and Models in Economic Dynamics</u>, (North Holland Publishing Company, Amsterdam).

Hordijk, L. and J.H.P. Paelinck, 1974, <u>Spatial Econometrics : Some Contributions</u>, Netherlands Economic Institute, Foundations of Empirical Economic Research, no. 6.

Hodgman, C.D., 1961, <u>Standard Mathematical Tables</u>, Chemical Rubber Publishing Company, Cleveland, (12th ed., 4th printing).

Hordijk, L. and J.H.P. Paelinck, 1975, Spatial Econometrics : Some Further Results, Netherlands Economic Institute, Foundations of Empirical Economic Research, no. 5.

Isard, W. and P. Liossatos, 1975, "Parallels from physics for space-time development models", part 1, Regional Science and Urban Economics, Vol. 5, no. 1, pp. 5-40.

Klaassen, L.H. and A.C. Van Wickeren, 1969, "Interindustry Relations : an Attraction Model, a progress report", in H.C. Bos (ed.), Towards Balanced International Growth, (North Holland Publishing Company, Amsterdam).

Leontief, W. and A. Strout, 1966, "Multiregional Input-Output Analysis", in W. Leontief (ed.), Input-Output Economics (Oxford University Press, New York), pp. 223-257.

Meriam, J.L., 1971, Dynamics, (Wiley, New York, second edition).

Paelinck, J.H.P., 1973, Modèles de politique économique multi-régionale basés sur l'analyse d'attraction, L'Actualité Economique, octobre-décembre, pp. 559-564.

Paelinck, J.H.P., 1974, "Een aanloopje tot het gebruik van tensor-analyse bij het opzetten van economische milieu-modellen", in P. Nijkamp (ed.), Milieu en Economie, (Universitaire Pers Rotterdam, 1974).

Strauss, A., 1968, An Introduction to Optimal Control Theory, (Springer Verlag, Berlin).

Theil, H., 1971, Principles of Econometrics (North Holland Publishing Company, Amsterdam).

INTERREGIONAL ATTRACTION THEORY : A GENERALISATION
OF ATTRACTION THEORY THROUGH INTERREGIONAL INPUT-OUTPUT

D. Van Wynsberghe
Regional Economic Council for Brabant
Brussels - Belgium

I. Introduction

The development of the attraction theory as first introduced by Professor L.H. Klaassen and later also by A.C. Van Wickeren can be seen as an extension of standard input-output analysis under simplified conditions on the technical coefficients and the allocation coefficients. The equality of demand and supply inherent to the Leontief approach is disregarded in attraction theory so that Professor Klaassen's merit can be summarised as follows : "Leontief's demand model is stated in a much wider framework, in which demand and supply come to their own right". (Van Wickeren, 1972).

It is the aim of this paper to reexamine the attraction theory in a closed interregional system and to see to what extent both approaches could achieve identical results. Anticipating on what follows we can state that although attraction theory is a generalisation of standard input-output theory, it is at the same time a simplified view on interregional input-output with the condition that the own region comes first. The consequences of the interregional dimension for the regional growth are examined to some extent in order to arrive at some general conclusions on the interaction of long range development of interregional growth, resource shortages and pollution.

II. A Simple Version of the Attraction Model

The main hypotheses of attraction theory are the following : Communication costs (including transportation costs) are becoming more important and prevent economic activity from spreading arbitrarily over a given space so that the extent to which two sectors are present on the regional level can be seen as a measure of the degree of attraction between these two sectors.

In order to be as close as possible to attraction theory we will use the same symbols as A.C. Van Wickeren (1972) so that the model can be summarised as follows :

Let it be assumed that the communication costs affecting the location of industry "k" in the (relevant) region "j" are :

$$_jT_k = t_{kd}(_jg_k - _jd_k) + \Sigma t_{hk}(\beta_{hk}\, _jg_k - \alpha_{hk}\, _jg_h) \quad (1)$$

$$(h = 1,\ldots,t)$$

The symbols in equation (1) have the following meaning

- $_jd_k$: Total output of sector k in region j;
- $_jT_k$: Total communication costs affecting the location of industry "k" located in region "j";
- t_{kd} : communication costs, exposed by industry "k" in order to "export" one unit of its product to other regions;
- $_jd_k : \Sigma_h\, _jr_{kh} + _jf_k$: total of intermediate and final demand for products "k" in region "j";
- t_{hk} : communication costs involved in the "import" of one unit of product "h" by industry "k";
- $\beta_{hk}\, _jg_k$: requirements of industry "k" located in "j" for intermediate products "h" defined by technical coefficient β_{hk}
- $\alpha_{hk}\, _jg_h$: amount of intermediate products "h" produced in region "j" and available for industry "k" in "j" defined by allocation coefficient α_{hk}

It is clear that according to this "definition" only communication costs on interregional import and exports are considered and that we must be in a situation so that

$$_jg_k \geq\, _jd_k \quad (2)$$

$$\beta_{hk}\, _jg_k \geq \alpha_{hk}\, _jg_h \quad (3)$$

As we want to give a simplified version which permits a quick insight in the nature of the theory, the conditions that all t_{kd}'s and t_{hk}'s are equal and that each region is completely self supporting in the field of communication facilities will be introduced. They do not influence the results. This simplification can be removed without calling for new theoretical problems.

The solution of the model is obtained by setting :

$$_jT_k = \alpha_{tk}\, _jg_t \text{ in (1) so that after some manipulations} \quad (4)$$

and rearrangements the reduced form

becomes : (t stands for the communication sector)

$$_jg_k = \frac{t_{kd}}{t_{kd}+\Sigma t_{hk}\beta_{hk}} \,_jd_k + \Sigma \frac{t_{hk}\beta_{hk}}{t_{kd}+\Sigma t_{hk}\beta_{hk}} \cdot \frac{\alpha_{hk}}{\beta_{hk}} \cdot \,_jg_h$$

$$+ \frac{\beta_{tk}}{t_{kd}+\Sigma t_{hk}\beta_{hk}} \cdot \frac{\alpha_{tk}}{\beta_{tk}} \cdot \,_jg_t \tag{5}$$

Equations (5) can be rewritten as :

$$_jg_k = \lambda_{kd} \,_jd_k + \sum_{l=1}^{n+1} \lambda_{lk} \cdot \frac{\alpha_{hk}}{\beta_{hk}} \,_jg_l \tag{6}$$

with n sectors and as the (n+1)th sector the communication sector.

If in (6) all $\lambda_{lk} = 0$ than the system can be reduced to the standard input-output system by the use of intermediate demand, the technical coefficient and final demand. The second part of the right hand side of (6) adds a supply element to the already existing demand element in the standard Leontief approach.

III. Re-Examination of the Attraction Model

1. The regional demand.

The definition of the communication cost $_jT_k$ seems to be at the origin of interesting results (5) or (6).
Indeed what is supposed is that exports out of the region are counted for an amount $(_jg_k - {}_jd_k)$ which means that no production k of other regions is brought into the region j as long as there are products k produced in j available.
In other words the own regional production of a given sector goes by preference to the own regional demand for goods of that sector.
This is a very "heroic" hypothesis that does not permit any product mix in the sectors nor between the regions and finally and practically reduces the whole regional economies to some 50 to 100 homogenous sectoral outputs. It seems to us that what is gained by introducing the spatial dimensions is lost by the product homogenity constraint. How can this constraint be weakened ?

If we accept that just as has been done in (1) for the intermediate inputs, output also is only for a certain fraction available on the local (regional) market then (1) becomes :

$$_jT_k = t_{kd}(_jg_k - _j\gamma_k \cdot _jd_k) + \Sigma t_{hk}(\beta_{hk}\, _jg_k - \alpha_{hk}\, _jg_h) \qquad (7)$$

The reduced form of (7) becomes as in (5)

$$_jg_k = \frac{t_{kd}}{t_{kd}+\Sigma t_{hk}\beta_{hk}} \cdot _j\gamma_k\, _jd_k + \Sigma_h \frac{t_{hk}\beta_{hk}}{t_{kd}+\Sigma t_{hk}\beta_{hk}} \cdot \frac{\alpha_{hk}}{\beta_{hk}} \cdot _jg_h$$

$$+ \frac{t_{tk}}{t_{kd}+\Sigma t_{hk}\beta_{hk}} \cdot \frac{\alpha_{tk}}{\beta_{tk}} \cdot _jg_t \qquad (8)$$

A comparison of (5) and (8) shows that in both cases $_jg_k$ is explained by $_jd_k$, $_jg_h$ and $_jg_t$ but that the weighting coefficient differs. Supply of the $_jg_h$ does not change as it was already introduced for only that part that was available on the local market, but demand $_jd_k$ is reduced to its real value $_j\gamma_k\, _jd_k$ (competitive market).

2. The interregional relations

The attraction theory uses the concepts : technical coefficient and allocation coefficient. The first one is obtained by looking at the columns of the I.O. table and the second only by looking at the rows of the table.

If we accept that regions are defined in a "proper" way, then we will show that this distinction can be dropped and the interregional structural coefficient can replace both.

We accept the following definitions for a 3 region version (Table 1) (Van Wynsberghe, 1974).

Table I - Interregional Input-Output (9)

	Destination / Region of origin	Intermediate demand			Final demand			
		Region 1	Region 2	Region 3	Region 1	Region 2	Region 3	Total
		intermediate sector	intermediate sector	intermediate sector	final sector	final sector	final sector	
sectors	Region 1	${}^{1}_{i}v^{1}_{j}$	${}^{1}_{i}v^{2}_{j}$	${}^{1}_{i}v^{3}_{j}$	${}^{1}_{i}f^{1}_{k}$	${}^{1}_{i}f^{2}_{k}$	${}^{1}_{i}f^{3}_{k}$	${}^{1}_{i}q$
	Region 2	${}^{2}_{i}v^{1}_{j}$	${}^{2}_{i}v^{2}_{j}$	${}^{2}_{i}v^{3}_{j}$	${}^{2}_{i}f^{1}_{k}$	${}^{2}_{i}f^{2}_{k}$	${}^{2}_{i}f^{3}_{k}$	${}^{2}_{i}q$
	Region 3	${}^{3}_{i}v^{1}_{j}$	${}^{3}_{i}v^{2}_{j}$	${}^{3}_{i}v^{3}_{j}$	${}^{3}_{i}f^{1}_{k}$	${}^{3}_{i}f^{2}_{k}$	${}^{3}_{i}f^{3}_{k}$	${}^{3}_{i}q$
primary inputs	Region 1	${}^{1}_{p}b^{1}_{j}$	${}^{1}_{p}b^{2}_{j}$	${}^{1}_{p}b^{3}_{j}$				
	Region 2	${}^{2}_{p}b^{1}_{j}$	${}^{2}_{p}b^{2}_{j}$	${}^{2}_{p}b^{3}_{j}$				
	Region 3	${}^{3}_{p}b^{1}_{j}$	${}^{3}_{p}b^{2}_{j}$	${}^{3}_{p}b^{3}_{j}$				
		q^{1}_{j}	q^{2}_{j}	q^{3}_{j}				

with for a general symbol y

${}^{r}_{i}y^{s}_{j}$ from sector i to sector j and from sector r to region s

- with for the output :

$$W_{(3n \times 1)} = \begin{bmatrix} {}^1_i q \\ {}^2_i q \\ {}^3_i q \end{bmatrix} \quad \text{and} \quad \hat{W}_{(3n \times 3n)} = \hat{W}^1 \quad (10)$$

and $_l$ a vector with units of appropriate dimension

- with for the intermediate sectors :

$$V_{(3n \times 3n)} = [V^1; V^2; V^3] \quad (11)$$

and

$$V^3_{(3n \times n)} = \begin{bmatrix} {}^1 V^s \\ {}^2 V^s \\ {}^3 V^s \end{bmatrix} \quad (12)$$

so that :

$$V \cdot \hat{W}^{-1} = \begin{bmatrix} r_\beta{}^s \\ i \; j \end{bmatrix} \quad \text{interregional structural} \quad (13)$$
$$(3n \times 3n) \quad \text{coefficients}$$

- with for final demand for products per region of origin :

$$F_{(3n \times 3m)} = \begin{bmatrix} {}^1 F \\ {}^2 F \\ {}^3 F \end{bmatrix} \quad (14)$$

and with

$$r_F{}_{(n \times 3m)} = [r_F 1; r_F 2; r_F 3] \quad (15)$$

and

$$\overset{\star}{F}{}^s = l' \begin{bmatrix} {}^1 F^s \\ {}^2 F^s \\ {}^3 F^s \end{bmatrix} \quad \text{the final demand per} \quad (15\text{bis})$$
$$\text{region of destination}$$
$$(1 \times 1)(1 \times 3n)(3n \times m)$$

with $\left[{}_i^r\gamma_k^s \right] = \begin{bmatrix} {}_1F^s \\ \hline {}_2F^s \\ \hline {}_3F^s \end{bmatrix} \cdot \hat{\vphantom{F}}_F^{*-1}$ the budget shares per   (15ter)
region of origin for
the final demand in the
region of destination

(3nxm)   (3nxm)   (mxm)

given that the allocation coefficients are defined by

$$V' \hat{W}^{-1} = \left[{}_i^r\alpha_j^s \right] \qquad (16)$$

(3nx3n)

and by (13) that $[\beta]' = \hat{W}^{-1} V'$ or $[\alpha]' = \hat{W}^{-1} V$

so that $\hat{W}[\alpha]' \hat{W}^{-1} = [\beta]$ what means that   (17)

$[\alpha]'$ is a similarity transformation of $[\beta]$ and also that

$$\hat{W}[\beta]' \hat{W}^{-1} = [\alpha] \qquad (18)$$

Under the implicit conditions of this paper, the use of an $[\alpha]$ matrix can easily be overcome and the $[\beta]$ matrix includes all information needed for the intermediate sectors. The $[\gamma]$ matrix with the budget shares will take care of final demand.

3. The communication cost structure

We introduce the following definitions :   (19)

       ${}_i^r t_j^s$ the unit communication cost in region r to export one unit of sector i to sector j in region s (j can be replaced by k for final demand)

       ${}_j^s m_i^r$ the unit communication cost in r to import   (20) one unit from s for sector i from sector j (i can be replaced by k for final demand)

The total interregional communication cost is then given by the sum of :

ϕ_1 : the interregional exports costs

$$\phi_1 = \sum_{\substack{s \\ s \neq r}} \sum_{\substack{\forall j \\ \forall k}} {}^r_i t^s_j \left[{}^r_i \beta^s_j \cdot q^s_j + {}^r_i \gamma^s_k \overset{\star}{f}{}^s_k \right] \qquad (21)$$

ϕ_2 : the interregional import costs

$$\phi_2 = \sum_{\substack{s \\ s \neq r}} \sum_{\forall j} {}^s_j m^r_i \; {}^s_j \beta^r_i \cdot q^r_i + \sum_{s \neq r} \sum_{\forall k} {}^s_i m^r_k \; {}^s_i \gamma^r_k \overset{\star}{f}{}^r_k \qquad (22)$$

If we now define $\phi = \phi_1 + \phi_2 = \delta^r_i q^r_i$ with δ^r_i the average communication cost, then a result as in (3) and (8) can be derived but in this way that in the reduced form q^r_i is a function first of the final demand in all other regions of destination $\overset{\star}{f}{}^s_k$ and in the own region $\overset{\star}{f}{}^r_k$ and secondly of the output of the sectors of the other regions q^s_j.

$$q^r_i = \sum_{s \neq r} \sum_{\forall j} \frac{{}^r_i t^s_j}{\mu^r_i} \cdot {}^r_i \beta^s_j \, q^s_j + \sum_{s \neq r} \sum_{\forall k} \frac{{}^r_i t^s_k}{\mu^r_i} \cdot {}^r_i \gamma^s_k \overset{\star}{f}{}^s_k$$

$$+ \sum_{s \neq r} \sum_{\forall k} \frac{{}^s_i m^r_k}{\mu^r_i} \cdot {}^s_i \gamma^r_k \overset{\star}{f}{}^r_k \qquad (23)$$

with
$$\mu^r_i = \left[\delta^r_i - \sum_{s \neq r} \sum_{\forall j} {}^s_j m^r_i \; {}^s_j \beta^r_i \right] \qquad (24)$$

so that μ^r_i stands for the average export communication costs for the production of q^r_i and is an <u>implicit function</u> of the output of all sectors in the region r that contribute to the production of q^r_i.

So (23) can be rewritten as

$$q^r_i = \sum_{s \neq r} \sum_{\forall j} {}^r_i \lambda^s_j \; {}^r_i \beta^s_j \, q^s_j + \sum_{s \neq r} \sum_{\forall k} {}^r_i \lambda^s_k \; {}^r_i \gamma^s_k \overset{\star}{f}{}^s_k \qquad (25)$$

$$+ \sum_{s \neq r} \sum_{\forall k} {}^s_i \varepsilon^r_k \; {}^s_i \gamma^r_k \overset{\star}{f}{}^r_k$$

If we compare (23) to a static interregional I/O table

$$q_i^r = \sum_s \sum_{\forall j} {}_i^r\beta_j^s q_j^s + \sum_s \sum_{\forall k} {}_i^r\gamma_k^s f_k^{x_s} \quad (26)$$

Then it follows that the main difference lays in the fact that the regional output of other sectors q_j^r is not <u>explicitly</u> present in (25).

If (25) is estimated on a time series then the λ's and ε's permit to draw conclusions on the relative magnitudes of the communication costs ${}_i^r t_j^s$, ${}_i^r t_k^s$ and ${}_i^s m_k^r$.

If we introduce a two region version with a homogeneous output per sector, then formulas (21) and (22) can be reduced to the regional attraction model as developed from (1) to (5) because of the fact that all (γ) are zero by definition so that export can be defined as output minus total regional demand. The interregional structural coefficients ${}_j^s\beta_i^r$ can be replaced by the regional technical coefficients minus the own available inputs and a transformation of the type of formula (17) can bring in the allocation coefficients and the regional outputs of the other sectors.

IV. Consequences for Interregional Growth and Development

The introduction of the interregional input-output conception in the existing framework of attraction theory leads to some interesting results.

1. Interregional competition and development

The output of a given sector, in a given region is now no longer a function of the regional variables alone but a direct function of the outputs and the final demands in the relevant sectors of <u>the other regions</u> and an indirect function of the own regional output of the other sectors. This realistic vision and the fact that the regional output of a certain sector is not of a homogeneous nature in space and that relative competitive advantages do exist are reflected in interregional attraction theory through a reduction of outputs and demand to the <u>competitive quantities only.</u>

This means that regional sectoral expansion is not in the first place a function of the communication costs for exports and imported products and of the own regional demand but of the already existing situations in the other regions (production, demand and communication costs).

2. Economic regions and regional economic science regions (RES Regions)

Economic regions in the sense of the interregional theory have to be defined so that the unit communication costs within the regions are constant at a given moment and for each sector.

This means that the regions are different from sector to sector. A planned output in a given point in space <u>at a given moment</u> divides space in a number of regions for which the communication costs are more or less constant per unit (δ_i^r = constant).

So in this view, there are no uniquely defined "economic" regions. The question of how to define an economic region is without practical interest at the level of a given sector only.

The regional dimension - or the spatial dimension - in the "regional economic science" sense becomes important only if we are superposing all the theoritical regions defined by the pure communication cost criterium. By the use of constant average communication costs clusters of regions will appear in which sectors exist "near" to each other and form spacially a "hard core". The border regions however will be more favourable for some sectors and less for most others. They form the "corridors" of interregional expansion. Attraction in these border regions is great for certain sectors and as intersectoral competition is less than in the "hard core" a spatial expansion is generated, which is typical for the industrial axes.

So geografical or more general speaking ecological factors together with economical determinants play an important role in the interregional development pattern. Availability of natural ressources and access of the region can have a significant influence on the communication costs for a certain sector which may have a fast development in that region if its output can be competitive on other regional markets.

The regional definitions are no longer fixed but varying over time with changing conditions. The introduction of a new sector (or a non existing sector at the regional level) changes the regions as a function of the relative changes in the communication costs in all points in space. The change is reflected by the changes in market shares that are significant for the competitiveness of each sector in each of the regions.

Although such a vision on economic regions is interesting to explain
the interregional dynamic evolution, it is not very practical, indeed
almost impossible to use. Therefore a regional definition that is
operational and constant with the developed theory could be the
following one.

A region in the regional economic science (RES) is composed of those
clusters of points in space which have similar changes in average
growth rate of their economic activity. So by time series analysis
RES regions can be defined. We want to stress the importance of the
word "changes" or variations in average growth rate instead of the
average growth rate criterion. Indeed a sub-urbain area or the
hinterland of a growth pole does not have the same average growth rate
as the growth pole itself but - if it is really mutually interacting
with the growth pole - it must have a similar variation in its growth
rate as the growth pole.
The criterion for a RES region is not the speed of economic development
but the degree of acceleration of economic development. It should be
clearly understood that the used criterion will explicitly rule out
those points which are under a one-way influence (economic domination)
of a growth pole.

It can be assumed that the market share and budget ratio's in RES
regions are rather stable in the short and medium term. The long run
variations will be the rule.
This is due to the underlying concept of economic regions or superposition of economic regions as developed before.

3. Long range developments of regional economic growth

Two major facts are important for communication costs. First their
absolute magnitude and secondly their relative variations over time
and space.

High communication costs - due to e.g. a limited technical knowledge,
or to infrastructural problems - but which do not differ much in space
will lead to a regional development of the type described by E.M. Hoover
and know as the "stage theory". (North, 1970)

If on the other hand, communication costs are spatially very different
for a number of sectors then a development as described by D.C. North
and known as the "export base" theory may occur. (Stabler, 1968)

If communication costs between a number of sectors are high and can only be reduced by a spatial grouping of specific sectors, then a regional development as first described by F. Perroux and later on developed by J. Boudeville and others in the well known "growth pole" theory will occur. Growth will not appear everywhere but in certain points or poles through the effect of "external economies" in the Marshall sense. There will be different growth rates for different sectors, sectors will appear and others disappear so that structural change will be the rule. Hirschman postulated that economic development must go through such growth poles.

So in analysing the past, and if "lessons from the past" are meaningful for the future, then the regional differences in communication costs could be growing together with the concentration of the economic activities so that the developing world has no hope to reallocate economic activity by means of the existing economic mechanisms.

Can marginal communication costs in a hihly industrialized region become higher than marginal communication costs in a developing region ? Do there exist already sectors for which communication costs are lower in less developed regions ? Will resource shortages and pollution problems reserve the existing trend ?

And if the answer is yes for some sectors, will or can these sectors give the start to a new regional development of the developing world ? A positive answer is doubtfull because of the true nature of these problems : they are universal and absolute problems. Resource shortages will be felt in most of the regions, pollution cannot be solved at a regional level but must be seen at a world level.
Transferring polluting industries to developing area's is in fact no solution of the ecological problem in the long run. The solution to resource shortages and pollution must be found in new technologies and integration of the production process in the general ecological process. This means that heavy investments will have to be payed by the developed world. So most probably resource shortages and ecological problems will not redistribute income between developing and developed regions unless the developed regions change their preferences. The poor cannot become richer if the rich do not become less eager on the expected profits. Someone has to pay the price.

V. Conclusions

The following main conclusions can be drawn this contribution:

(1) Standard input-output is a simplified version of attraction theory. This last theory operates under a simplified version of interregional input-output: the own region has an absolute preference for the own regional production and there is only one homogenous product per sector.

(2) Interregional attraction theory puts regional development in an interregional framework. So the output of a certain sector in a certain region is no longer a function of regional variables alone. The economic variables in other regions do directly influence the regional output.

(3) In this view and under a free market system there will be no autonomous development in a region as long as it is not competitive with its output on other regional markets or on its own market (in competition with the imports from other outputs).

(4) The degree of competition of a regional output is reflected in its different market shares and budget shares which in turn are functions of the communication costs.

(5) An <u>economic region</u> has to be defined at the level of a given sector for a cluster of points in space and time where unit communication costs are constant (δ_i^r). Such an economic region is not uniquely defined and cannot be used as such for interregional attraction theory nor for interregional input-output.

(6) A region in the regional economic science sense (RES Region) is obtained by superposition of the different sectoral economic regions for a cluster of points in space. It consists of a "hard core" where unit communication costs should be fairly equal and of border regions which are favourable for a limited number of sectors and form the corridors of interregional economic expansion. RES regions are <u>practically</u> defined as clusters of points in space where similar <u>changes</u> in average growth rate appear over time.

(7) Interregional attraction theory is consistent with each of the existing regional development theories depending on the hypothesis concerning the communication costs e.g. export base theory, development stages theory and growth pole theory.

(8) If the regional differences in communication costs are growing as in the past when concentration of the economic activities goes further on, then the developing countries have no hope to reallocate economic activities with the existing economic mechanisms.

(9) The consequences of universal problems such as ecological problems, pollution and resource shortages will not necesseraly alter the existing trend. They can only result in the reallocation of income if they can start an economic development in the developing regions. This is - as suggested by the interregional attraction model - a function of the conditions in all the regions. So these universal problems will only change the conditions if they are felt differently in the long run in the developing and developed regions. One can doubt if this is the case.

References

North, D.C., 1970, "Location theory and regional economic growth" in Regional Economics of D.L. Kee, R.D. Dean and W.H. Lealy (Free Press, New York), pp. 30-31.

Stabler, J.C., 1968, Exports and evolution : the process of regional change, Land Economics, XLIV, No. 1, 11-23.

Van Wickeren, A.C., 1972, Interindustry relations, some attraction models (Rotterdam University Press, Rotterdam), 5-7.

Van Wynsberghe, D., 1974, "An operational non survey technique for estimating a coherent set of interregional I/O tables", Paper presented at the VIth International I/O Conference, Vienna.

NEW INTERNATIONAL ECONOMIC ORDER AND RESOURCE ALLOCATION IN DEVELOPING COUNTRIES

P.N. Mathur
University College of Wales
Aberystwyth - England

I. Introduction

Due to the politico-economic developments of the 1970's "the international economic order which provided a post-war generation of economic peace has collapsed, and we do not know what will take its place. East-West economic relations in a world of detent, rather than cold war, management of international economic structure in a world of three major power centres and the role of ever more asserting developing countries in the world economy .... A new crisis could severely disrupt the world economy and international economics until a new system is constructed to replace the Bretton Wood and GATT regimes, which have collapsed ... The hierarchy among nations is now uncertain, and the goals of major and many lesser countries often conflict, at least in the short run"[1]. The economic theory maintaining that free trade and investment potentially maximizes general economic welfare, has been based on, among other things, a tacit static assumption of trading partners remaining at more or less the same developmental stage, viz., their factor proportions do not drastically change, say through much faster capital accumulation in one trading country compared with the other. This theory got credence from the experience of stability of the past orders, based on the 'gold standard' under the leadership of the United Kingdom for many years, or a Bretton Woods regime operating under the economic dominance of the United States. The basic concepts underlying international economic policies have remained intact for an impressive period of time because, among other things, there was a politically accepted (or enforced) economic leadership which effectively ensured that the proportionate factor endowments of different participants in trade did not materially get altered, and which in effect relegated most of the underdeveloped world to a capital-scarce - labour/land abundant category. Change in the political situation has now made this world view increasingly difficult to maintain.

This whole scenario is a transfer on an international scale of our experiences of sectoral growth in national sectors and its politico-economic repercussions on the national scene. It is a well-known fact of economic history that manufacturing sectors could make semi-monopolistic type of cartels, much earlier in the course of their development than agricultural and similar ones. The total market phenomena then became where some sectors had got a monopolistic power, while others were selling their goods in a competitive framework. This obviously allowed the former to appropriate for themselves most of the gains from trade among them. During the early days of British industrialization it was soon realised that the latter could easily recover their terms of trade, if their commodities could be made really scarce. Hence came the Corn Laws. This politico-economic power of agriculturists was eroded after the repealing of the Corn Law, 1846, which allowed imports of agricultural products from the U.S.A. and other countries. Slowly, the same power is now being felt through the development of effective agricultural lobby in one country after another, showing its full development in the current U.S. and E.E.C. agricultural policies.

Similarly, on the international scale, early industrialising nations could get the advantage of their economic and political organisations in the determination of the terms of trade between themselves and raw material producing countries. With the politico-economic developments of recent decades, the whole balance is changing in the same way as it did in the national economies, and the new economic order must take account of the fact of effective economic semi-monopolistic powers being available to quite a few groups of producers apart from those producing advanced manufactures.

The rate of capital accumulation of various countries and the stage of their technological development has to become an integral part of the international trade and investment scenario, together with factor endowments, etc. Most countries have realised that they are much more policially independent and are now able to pursue far more independent economic policies. Quite a few politicians intend to use this newly acquired economic power in furtherance of their objective to achieve as quickly as possible transformation of their country into a modern developed economy. Thus, many governments have recently been deviating increasingly from the policy prescriptions of the traditional model, revealing a preference for reduced - rather than greater - freedom of trade and factor movement. Governments are clearly groping because nothing has been developed to replace the traditional theory guiding behaviour in details. If nothing does replace it, policy will be determined ad hoc - and tend to be dominated by whatever special interest can master the most political power at the proper strategic moment.

If the above understanding of the situation even partly corresponds to reality, it is clear that international economic policies, to be generally acceptable and feasible in a world of non-hierarchical comity of nations, with multi-centres for economic decisions, should ensure that in such a system no country feels that the system is giving it a short shrift and is an impediment for its relative growth and thus may have an incentive in disrupting the system or not cooperating with it. This could also encourage a rapid technological transformation (modernisation) of the economy of all the trading partners whose labour force is still using obsolete equipment - without not only harming any other participating country, but also ensuring that new innovations have full advantage of a free international economy to be able to come to full fruition, so as to continue to the process of growth in developed countries. International trade policy may have to adjust its role to serve these overall requirements. These requirements are quite vague, and, therefore, some rough and ready measure of equitable justice may have to be agreed upon. There is no simple theoretical or operational framework available at this time which will ensure such achievement by the world community. The purpose of the present paper is to try to develop a computable model of handle this problem.

II. Organisational Setup

For maximising economic growth and/or welfare, the following types of organisations have been successfully tried in the past :

(1) Free Trade : This led to the growth of Great Britain to a modern economy, and also showed a stability of the past orders extensibly based on the gold standard, under the clear leadership of the United Kingdom. This type has been very helpful for developed countries, as well as primary development of 'enclave economics', which could provide some necessary raw material or consumer good to the markets of the developed countries.

(2) Protective Regimes : Under this form countries like the U.S.A., Germany, the U.S.S.R., Japan, etc., could become developed economies, in spite of the better competitive power of the United Kingdom at these times. Through finely adjusted tariff walls around their economies

they could ensure that the internal market for their developing manufactures is reserved for the products of their own country, and thus they could uncouple the price structure of their countries with the international market and see that their internal price structure is the one suitable for the development of their own industries, while they could partake in international trade profitably at the international price-structure.

(3) Common Market Strategy : This has been adopted successfully first by advanced West European countries after the war. In this they had a common tariff (sometimes also quota restrictions) wall for the outside world, while the trade within the community was free. This proved to be quite beneficial for advanced economies which were more or less at the same stage of development, by giving them a big internal market for the development of new large-scale industries, as well as better competition to ensure that unhealthy firms did not survive in a small monopolistic market. And it also gave them protection from the outside world - both of developed countries like the U.S.A., and of developing countries who were now entering into primary manufacturing stage.

(4) Pressure Groups of Commodity Producing Countries : These types of groupings have been tried for long under various institutional arrangements, to keep up the prices of various commodities that prominently figured in international trade. However, the remarkable success of this grouping has only now been demonstrated by the Organisation of Petroleum Exporting Countries (OPEC), which was able to raise their intake from petroleum almost four-fold in two years' time. These types of economic organisations, with their separate decision centres in producing countries, will also have their effect on the working of the world economy and should be taken into account in the medium term future.

(5) Non-Market Systems : Apart from these, there are success stories of detailed planned non-market communist economies which have used international trade almost on a barter basis. There is no simple method of price determination within these economies. They may have to be treated in a slightly different way from the market economies.

As far as developed market economies are concerned, the pattern of organisation set by the EC and GATT seems to be the model for the near future. Among themselves these countries will compete, bargain and decide major economic issues through summitary. Their rate of growth, as a whole will depend not only on the rate of growth of their labour force, but also on the share of fixed investment in the CDP, and this share itself will be a function of their terms of trade with the developing countries, ad well as of the defence and other non-poductive expenditure. It may be noted that most of the innovations are expected to occur in this group.

For organising their developmental efforts, getting inspiration from the successful EC experiment, developing countries have already shown a tendency of making sub-groups among themselves. However, so far those experiments have not been very successful. One reason seems to be that within the grouping the provision of free trade meant undue advantage for some of the participating countries. And further, the establishment of individual major projects gave apprehensions in the minds of participants of further differential advantages likely to be reaped by the fortunate few. Thus, the likely pattern to emerge may be that under the cover of an overall insulation of the regional market from the world economy, individual country markets within the region may further be partly insulated from each other by well-designed, and smaller, tariff walls. Then, the individual countries within a region may develop in the same way as Germany and the U.S.A. developed in the world economy in the last century, viz., behind the protection of the tariff wall with countries more advanced than them within the region, while the region as a whole may partake into the advantages of a bigger market for the agreed commodities. This will mean a different regional price structure from international prices (developed countries' price system), as well as a slight insulation of the price structure of individual countries, in such a way as to allow equal benefit of growth to different countries within it. This further implies a separate monetary management (and probably unit) for transactions within such regions.

The development of developing countries depends on the adoption of new or modern technology for producing all the economic goods. It is not necessary that the total transformation (in individual industries) should be done at a single stroke. Usually some part of the final processing is first adopted, and then slowly more and more parts get manufactured in the developing economy itself. The economics of the production of each part in individual industries may be different. It is noticed that the countries with relatively easy access to foreign exchange do not tend to go much deeper into local production, and find it profitable to confine themselves only with the last stages, while importing semi-manufactured goods. As development proceeds and foreign exchange becomes scarcer and scarcer, it becomes necessary to manufacture more and more parts entering into the manufacture.

In a grouping we expect the countries to be in different stages of growth in individual industries. They together can programme for the maximum growth of the group as a whole, as well as individual countries, through detailed programming techniques[2]. This exercise will throw up the group's optimum relative price structure, as well as the tariff structures that should be followed within the group. This will, of course, be different from international one. It will be noted, therefore, that no industry by industry analysis will be able to allocate resources in a desirable way, because the international prices of inputs and outputs will not be suitable for that purpose. Further, in finding out an equitable benefit for individual countries through this process, it is not only necessary to get at an equitable rate of transformation of old technology to the new, but also to get equitable increases in current consumption of different economies. The structure of this consumption will depend upon individual preference functions. These will also interact with the internal price formation process discussed above.

Apart from these, we expect that commodity producers' groupings, like OPEC, will become a part of the world economic landscape. This would imply that the prices of primary commodities traded in world markets will be determined in accordance with the monopolistic principles rather than competitive ones which govern them at the moment. Thus the changes in supply and demand are likely to be accomodated by changes in production and/or stock management, rather than by the changes in the prices of the commodities, and the prices will be fixed in accordance with the demand schedule of the commodity rather than the supply considerations. And further, the suppliers will have to be careful not to increase the price of their commodity so much as to allow substantial development of its substitutes. They will also have to give sufficient attention to the development of that part of the international economy which will increase the demand of their commodity.

III. Mathematical Recapitulation of the Whole System

The wole new international economic order, described above, can be mathematically put as follows :

1) Commodity Balance

The basis balances connecting different regional models of the world economy will be International Commodity Balance Equations, as follows:

$$\sum_{R_1}^{R_1} D_{it}^1 (P_{it-1}) + \sum_{R_2}^{R_2} D_{it}^2 (P_{it-1}) = \sum_R S_{it} + B_{it} \qquad (1)$$

$$S_{it}^R = f_1(S_{it-1,t}^R, P_{it-1}, B_{it}) \qquad (2)$$

$$\frac{P_{it} - P_{it-1}}{P_{it-1}} = f_2(B_{it-1}) \qquad (3)$$

i=1,...,m, where there are m internationally traded commodities with managed markets.

Where D_i and S_i refer to demand and supply of ith commodity, supercript refer to region R_1 and R_2 beings sets of developed and developing regions. The assumption on the behaviour of these markets in the above formulation are that commodity supply is not affected by lagged price, as they will be determined by prior negotiations. B_{it} denotes the stocks of commodity i at time t.
As discussed above, it is further assumed that by the time the new system stabilises, all developing regions be grouped into common market type organisation, having their own programmes for development.

2) Developed Country Sub-Model

Following the conclusion of several studies done during the last decade or so, we shall take:

$$Y = \phi_1(\dot{L}, \frac{\Delta K}{Y}) \qquad (4)$$

Where Y, $\dot{L}$ and K are G.D.P., labour force and Capital stocks respectively, and a dot represents the rate of growth.

$$\frac{\Delta K}{Y} = \phi_2[(G.D.P.)_t, (G.D.P.)_{t-1}, (\text{Terms of Trade})_t, (\text{Balance of Payment on Current Account})_t] \qquad (5)$$

Here both G.D.P.'s are calculated in constant prices and short term effect of rises in import costs or that of export bonanza is also taken into account.

Further, the structure of investment in various industries is assumed to change systematically from year to year in response to changes in composition of final demand from the year t-2 to t-1.

This investment and its structure will determine the output and price pattern, given input-output coefficient, income determination equations and consumption patterns.

Once output and its pattern, as well as consumption and investment demand and their pattern, are determined, they can be used to derive import demand on both intermediate and final goods account. The second one, of course, will be directly a function of international commodity prices, while the first one will be indirectly so. These demands will be fed into the World Commodity Balance equations given above.

3) Developing Country's Sub-Model

We may assume that developing countries make their joint plans on a rather medium term basis for which the prices of internationally traded commodities are fixed. With appropriate stock management of commodity markets, they will get an indication of the availability of Free Foreign Exchange to them for a short planning period. They will also be able to plan the production of the commodities that they supply to the international market for the same period. With changing world demand they will have to revise their estimates of both periodically.

The coordination of a common market of developing countries does not mean the allocation of some major identifiable projects only. Apart from that, for becoming effective, allocation of hundreds of small industries may have to be thought of, otherwise the equitable distribution of direct and indirect advantages would become almost an impossible task. The decision to put small industries at any particular place is primarily taken at a decentralised level. Therefore, the guide-lines or signals should be avilable for them by a real or shadow pricing system. In the latter case the effective instruments for their generation and communication may have to be devised. Further, to ease trading relationship within the group is the very nature of a common market. An unrestricted trading relationship, as we have seen before, may lead to advantage of one country or another, unless and until there is a system of tariffs which equalises those advantages. However, if we have to have such a system, it has to be determined in accordance with some agreed principles. Those principles should be such as to lend themselves to the quantitative deduction of the tariff rate. If these motives are not previously agreed, to which yearly rates can be adjusted, there is a lot of scope for fundamental disagreements, which will, again and again, become bones of contention, as slight changes in one direction or another would be benefitting one group rather than the other. So, we not only want to determine the allocation of some identifiably big projects among different countries, but we also need to determine a criterion on which tariff rates between common market countries themselves can be based from time to time.

Further, as partly finished goods can be a marjor part of intra-regional trade, we should plan for processes rather than industries, as is done by multi-national companies or in European socialist countries. We can not define an industry like machine manufacturing or cemicals, or even textiles, when we know that a major portion of the inputs of these industries are the outputs of the same industries of some other country[3].

The presentation here, as a first step, is in a static setting with the usual assumptions of linear programming framework. A number of technical structures (as represented by technical coefficient matrices) are available to each of the countries. For the simplicity of exposition we shall assume that there are two such matrices : (1) Matrix A^*, representing an incomplete production structure, and (2) Matrix A, representing a complete one. There may be imports and imported inputs even in structure A, but these imports are competitive, and hence substitutable. Matrix A^* is further split up into two matrices, so that :

$$A^* = A^d + A^m \tag{6}$$

where A^d is domestic inputs matrix, and A^m is imported inputs matrix. Imports required by A^m are non-substitutable.

The basic balance relation between inputs, outputs, trade and final demand for country r can be expressed as :

$$(I-A^*)X^{*r} + (I-A)X^r + \sum_s X^{sr} - \sum_s X^{rs} + X^{wr} \leq Y^r \tag{7}$$

Where X^{*r} : Production vector under technology A^*

X^r : Production vector under technology A

X^{sr} : vector of imports from countries s to r

X^{rs} : vector of exports to countries s from r

X^{wr} : vector of imports from countries outside the common market (s, r=1,2,...,c, where c is the number of countries in the group, and s ≠ r)

Y^r : final demand + exports to outside world.

Where commodity exports to outside world are given by commodity model and others are estimated separately.

The restraining feature of the import based technology would be represented by another set of constraints :

$$(-A^m)X^{*r} + \Sigma X^{sr} + \Sigma X^{wr} > 0 . \qquad (8)$$

This constraint states that the total imports must at least equal the requirements of the import based technology.

There are certain sectors which have to be balanced nationally. These are : construction, electricity, transport (internal) and trade and services. For such sectors and imports have to be ruled out. Hence X^{sr} and X^{rs} vectors would not include these activities.

For primary sectors (i.e., agriculture, mining, etc.) as well as for nationally balanced sectors, it may be reasonable to assume that home technology is most suited, because the technology in these cases very much depends on natural endowments and local conditions. Furhter, an upper limit on the production of natural resource based industry has to be imposed in the short run[4]. This constraint would appear as :

$$X_i^{*r} \text{ (or, } X_i^r) \leq \bar{X}_i^{*r} \text{ (or, } \bar{X}_i^r) \qquad (9)$$

where $\bar{X}_i^r$ is the upper limit on the production of primary sector output i (i=1,2,...,k, primary sectors).

Now the whole model ties up when we impose a trade balance constraints, as :

$$\Sigma_i X_i^{wr} + \Sigma\Sigma_{si} X_i^{sr} - \Sigma\Sigma_{si} X_i^{rs} = X^{rw} \qquad (10)$$

where X^{rw} is the share of country r in community foreign exchange earned through trading with countries outside the market.

Subject to the above four sets of constraints, we have to minimise the sum of investment necessary for achieving the additional equitable final demand in the countries of the region. This gives the following objective function :

$$\text{Minimize } Z = \underset{ri}{\Sigma\Sigma} (k_i^{\star r} X_i^{\star r} + k_i^r X_i^r) \qquad (11)$$

Where $k_i^{\star r}$ is the investment necessary to increase the capacity of ith incomplete capacity in r^{th} region by one unit, and k_i^r is the investment necessary to increase the same capacity with a complete technology. $X_i^{\star r}$ and X_i^r have the same meaning as above.

The solution of the above model will give us the extent of capacity that is to be created for production of each commodity in each region separately for each technology. Further, the dual of the above will give us the direct and indirect capital costs of satisfying each constraint. Let p^r be the dual variable relating to the balance equation (7) above. Then they will represent the direct and indirect capital cost of adding one unit of commodity i to the final demand of region r. Further let R_i^r be the dual value relating to the set of equations (9). Then they represent the rental value of the particular primary resource. However, here the rent is equal to the direct and indirect savings of capital because of the availability of a particular primary resource. Further, the dual value due to constraint (10) (E^{rs}), represents the difference in direct and indirect capital costs between the regions. This difference arises because the programme insists on balancing the trade, and its net effect would be a different effective rate of exchange between the countries of the region. The next section gives the results of an illustrative example.

IV. A Trial Developing Countries Sub-Model

This model has been applied for allocation of industries for an assumed common market, having input-output relations as that of India of 1963, of Malaysia of 1965 and of Sri Lanka of 1965. Only the conclusions are given here.

All the increases in production recommended by the exercise for Malaysia and Sri Lanka are for the technology which is already prevalent there. This implies that it seems best to go on developing final stages of production in these countries, and to import intermediate and capital goods for the time being. The necessary imports of capital and intermediate goods is to be done from India, where all stages of production are indicated.

It is further indicated that both Sri Lanka and Malaysia should export as much agricultural products as they can, highlighting their land abundance.

It envisages that only four activities out of thirteen should be additionally undertaken in Malaysia - viz., agriculture, basic metals, paper and printing and electricity generation. Similarly, it envisages four activities for Sri Lanka. They are, agriculture, basic metals, wood and wood products and electricity. While for India, it envisages additional production in all the thirteen activities.

To see how much this pattern of results is stable, we have used and alternative set of additional consumption and investment as a necessary final demand, and solved the problem all over again. The results of this second exercise were basically similar to the first, in the sense that it also did not envisage production in Malaysia and Sri Lanka by the other complete technique. However, in the case of Malaysia, it envisaged the additional production of two more commodities, viz., mining and wood products. And for India it envisaged no additional production of paper and paper products, as all of that could now be imported from Malaysia without disturbing the balance of payments.

This brings us to a very important consideration, and that is, that the pattern of final demand in individual countries will influence the final allocation of industries. Hence, in envisaging the allocation and trade pattern, we just can't look only to the comparative cost advantages in these short run programmings, but shall have to give due weightage to demand considerations. Thus, any simple cost-benefits type study looks to be completely out of place.

It might be recalled that the objective function in the exercise was extra capital investment required. Therefore, the marginal cost represents direct and indirect capital requirement of an unit of consumption in individual countries. The structure of this direct and indirect marginal capital cost was found to be identical in the three countries for all the tradeable commodities - it is different only in the case of electricity and basic metals. We have assumed that while electricity is not tradeable at all, there is no possibility of trade of basic metal products from Malaysia and Sri Lanka to India. This change of structure should affect all the pricing systems within the countries, unless and until, in an overall model, it is decided that the pricing system facing the rest of the world, and/or facing internal markets, should be so adjusted as not to be affected by this change in factor scarcities, through appropriate tariffs and subsidies.

Much more interesting is the marginal cost of balanced trade thrown up by the programme. It shows that the direct and indirect capital cost of importing from India for Malaysia is .73 per unit, which is compensated by the direct and indirect capital advantage of exporting to India by the same amount. This implies that the exports to India should be made more costly by an amount equivalent to the interest on this, while imports from India should be subsidies by an equivalent amount. Alternatively, we may say that the rate of exchange for trade with India and Malaysia should be adjusted so that this effect is taken care of. Similar remarks hold good for the adjustment of the rate of exchange between India and Sri Lanka, and Sri Lanka and Malaysia. With the current rate of exchange and free trade, the tendency will be for Malaysia and Sri Lanka to export to India without taking balancing imports. Further, it may be noted that in the alternative programme, with a different demand pattern we have got a different marginal capital cost structure, as well as a different direct and indirect capital cost of trading. While, between Malaysia and India, it is .73 in the first, it is only .65 in the term pricing pattern, as well as on exchange rate. Between India and Sri Lanka this cost is indicated as .87 per unit for both the programmes, between Malaysis and Sri Lanka it is .13 and .22 in the two programmes.

As we had envisaged while making the programme, there is a lot of comparative advantage in agricultural production in Malaysia and Sri Lnaka. In order to avoid giving the unrealistic rate of growth for those activities, we had put an upper limit for them. The result of that is the agriculture in both countries gives an extra rent element. It is .6 for Malaysia and .51 for Sri Lanka. Further, in the second one, mining also produced a rent component of .02 per unit in Malaysia.

This brings out clearly that for an effective common market, we require defacto different exchange rates within the common market countries than with the outside world if the countries going into economi cooperation are at different stages of economic development. This can only be derived through an overall economic analysis, taking into account all the economic factors - both on the supply and on the demand side.

Footnotes

1. See e.g. : 'The International Economic Order' by Bergsten, C.F. and Mathieson, J.A., Ford Foundation, New York, 1973, pp. 56-57.

2. An example of this problem has been elaborated by Mathur, P.N. "Industrial Location Strategy for a Common Market of Countries with Different Levels of Development and its Final Implications", (Forthcoming).

3. From this it follows that the basis data relating to industries of these countries will also have to be similarly disaggregated. The common tools like national income accounts, as well as input-output tables may have to be re-cast in requisite details.

4. A more elaborate discussion of this and some other points relevant to a regional model will be found in : Mathur, P.N. "Multiregional Analysis in a Dynamic Framework", in Input-Output Techniques (Eds. A. Brody and A.P. Carter), North-Holland, 1972.

Vol. 59: J. A. Hanson, Growth in Open Economies. V, 128 pages. 1971.

Vol. 60: H. Hauptmann, Schätz- und Kontrolltheorie in stetigen dynamischen Wirtschaftsmodellen. V, 104 Seiten. 1971.

Vol. 61: K. H. F. Meyer, Wartesysteme mit variabler Bearbeitungsrate. VII, 314 Seiten. 1971.

Vol. 62: W. Krelle u. G. Gabisch unter Mitarbeit von J. Burgermeister, Wachstumstheorie. VII, 223 Seiten. 1972.

Vol. 63: J. Kohlas, Monte Carlo Simulation im Operations Research. VI, 162 Seiten. 1972.

Vol. 64: P. Gessner u. K. Spremann, Optimierung in Funktionenräumen. IV, 120 Seiten. 1972.

Vol. 65: W. Everling, Exercises in Computer Systems Analysis. VIII, 184 pages. 1972.

Vol. 66: F. Bauer, P. Garabedian and D. Korn, Supercritical Wing Sections. V, 211 pages. 1972.

Vol. 67: I. V. Girsanov, Lectures on Mathematical Theory of Extremum Problems. V, 136 pages. 1972.

Vol. 68: J. Loeckx, Computability and Decidability. An Introduction for Students of Computer Science. VI, 76 pages. 1972.

Vol. 69: S. Ashour, Sequencing Theory. V, 133 pages. 1972.

Vol. 70: J. P. Brown, The Economic Effects of Floods. Investigations of a Stochastic Model of Rational Investment. Behavior in the Face of Floods. V, 87 pages. 1972.

Vol. 71: R. Henn und O. Opitz, Konsum- und Produktionstheorie II. V, 134 Seiten. 1972.

Vol. 72: T. P. Bagchi and J. G. C. Templeton, Numerical Methods in Markov Chains and Bulk Queues. XI, 89 pages. 1972.

Vol. 73: H. Kiendl, Suboptimale Regler mit abschnittweise linearer Struktur. VI, 146 Seiten. 1972.

Vol. 74: F. Pokropp, Aggregation von Produktionsfunktionen. VI, 107 Seiten. 1972.

Vol. 75: GI-Gesellschaft für Informatik e.V. Bericht Nr. 3. 1. Fachtagung über Programmiersprachen · München, 9.–11. März 1971. Herausgegeben im Auftrag der Gesellschaft für Informatik von H. Langmaack und M. Paul. VII, 280 Seiten. 1972.

Vol. 76: G. Fandel, Optimale Entscheidung bei mehrfacher Zielsetzung. II, 121 Seiten. 1972.

Vol. 77: A. Auslender, Problèmes de Minimax via l'Analyse Convexe et les Inégalités Variationelles: Théorie et Algorithmes. VII, 132 pages. 1972.

Vol. 78: GI-Gesellschaft für Informatik e.V. 2. Jahrestagung, Karlsruhe, 2.–4. Oktober 1972. Herausgegeben im Auftrag der Gesellschaft für Informatik von P. Deussen. XI, 576 Seiten. 1973.

Vol. 79: A. Berman, Cones, Matrices and Mathematical Programming. V, 96 pages. 1973.

Vol. 80: International Seminar on Trends in Mathematical Modelling, Venice, 13–18 December 1971. Edited by N. Hawkes. VI, 288 pages. 1973.

Vol. 81: Advanced Course on Software Engineering. Edited by F. L. Bauer. XII, 545 pages. 1973.

Vol. 82: R. Saeks, Resolution Space, Operators and Systems. X, 267 pages. 1973.

Vol. 83: NTG/GI-Gesellschaft für Informatik, Nachrichtentechnische Gesellschaft. Fachtagung „Cognitive Verfahren und Systeme", Hamburg, 11.–13. April 1973. Herausgegeben im Auftrag der NTG/GI von Th. Einsele, W. Giloi und H.-H. Nagel. VIII, 373 Seiten. 1973.

Vol. 84: A. V. Balakrishnan, Stochastic Differential Systems I. Filtering and Control. A Function Space Approach. V, 252 pages. 1973.

Vol. 85: T. Page, Economics of Involuntary Transfers: A Unified Approach to Pollution and Congestion Externalities. XI, 159 pages. 1973.

Vol. 86: Symposium on the Theory of Scheduling and its Applications. Edited by S. E. Elmaghraby. VIII, 437 pages. 1973.

Vol. 87: G. F. Newell, Approximate Stochastic Behavior of n-Server Service Systems with Large n. VII, 118 pages. 1973.

Vol. 88: H. Steckhan, Güterströme in Netzen. VII, 134 Seiten. 1973.

Vol. 89: J. P. Wallace and A. Sherret, Estimation of Product. Attributes and Their Importances. V, 94 pages. 1973.

Vol. 90: J.-F. Richard, Posterior and Predictive Densities for Simultaneous Equation Models. VI, 226 pages. 1973.

Vol. 91: Th. Marschak and R. Selten, General Equilibrium with Price-Making Firms. XI, 246 pages. 1974.

Vol. 92: E. Dierker, Topological Methods in Walrasian Economics. IV, 130 pages. 1974.

Vol. 93: 4th IFAC/IFIP International Conference on Digital Computer Applications to Process Control, Part I. Zürich/Switzerland, March 19–22, 1974. Edited by M. Mansour and W. Schaufelberger. XVIII, 544 pages. 1974.

Vol. 94: 4th IFAC/IFIP International Conference on Digital Computer Applications to Process Control, Part II. Zürich/Switzerland, March 19–22, 1974. Edited by M. Mansour and W. Schaufelberger. XVIII, 546 pages. 1974.

Vol. 95: M. Zeleny, Linear Multiobjective Programming. X, 220 pages. 1974.

Vol. 96: O. Moeschlin, Zur Theorie von Neumannscher Wachstumsmodelle. XI, 115 Seiten. 1974.

Vol. 97: G. Schmidt, Über die Stabilität des einfachen Bedienungskanals. VII, 147 Seiten. 1974.

Vol. 98: Mathematical Methods in Queueing Theory. Proceedings 1973. Edited by A. B. Clarke. VII, 374 pages. 1974.

Vol. 99: Production Theory. Edited by W. Eichhorn, R. Henn, O. Opitz, and R. W. Shephard. VIII, 386 pages. 1974.

Vol. 100: B. S. Duran and P. L. Odell, Cluster Analysis. A Survey. VI, 137 pages. 1974.

Vol. 101: W. M. Wonham, Linear Multivariable Control. A Geometric Approach. X, 344 pages. 1974.

Vol. 102: Analyse Convexe et Ses Applications. Comptes Rendus, Janvier 1974. Edited by J.-P. Aubin. IV, 244 pages. 1974.

Vol. 103: D. E. Boyce, A. Farhi, R. Weischedel, Optimal Subset Selection. Multiple Regression, Interdependence and Optimal Network Algorithms. XIII, 187 pages. 1974.

Vol. 104: S. Fujino, A Neo-Keynesian Theory of Inflation and Economic Growth. V, 96 pages. 1974.

Vol. 105: Optimal Control Theory and its Applications. Part I. Proceedings 1973. Edited by B. J. Kirby. VI, 425 pages. 1974.

Vol. 106: Optimal Control Theory and its Applications. Part II. Proceedings 1973. Edited by B. J. Kirby. VI, 403 pages. 1974.

Vol. 107: Control Theory, Numerical Methods and Computer Systems Modeling. International Symposium, Rocquencourt, June 17–21, 1974. Edited by A. Bensoussan and J. L. Lions. VIII, 757 pages. 1975.

Vol. 108: F. Bauer et al., Supercritical Wing Sections II. A Handbook. V, 296 pages. 1975.

Vol. 109: R. von Randow, Introduction to the Theory of Matroids. IX, 102 pages. 1975.

Vol. 110: C. Striebel, Optimal Control of Discrete Time Stochastic Systems. III. 208 pages. 1975.

Vol. 111: Variable Structure Systems with Application to Economics and Biology. Proceedings 1974. Edited by A. Ruberti and R. R. Mohler. VI, 321 pages. 1975.

Vol. 112: J. Wilhlem, Objectives and Multi-Objective Decision Making Under Uncertainty. IV, 111 pages. 1975.

Vol. 113: G. A. Aschinger, Stabilitätsaussagen über Klassen von Matrizen mit verschwindenden Zeilensummen. V, 102 Seiten. 1975.

Vol. 114: G. Uebe, Produktionstheorie. XVII, 301 Seiten. 1976.

Vol. 115: Anderson et al., Foundations of System Theory: Finitary and Infinitary Conditions. VII, 93 pages. 1976

Vol. 116: K. Miyazawa, Input-Output Analysis and the Structure of Income Distribution. IX, 135 pages. 1976.

Vol. 117: Optimization and Operations Research. Proceedings 1975. Edited by W. Oettli and K. Ritter. IV, 316 pages. 1976.

Vol. 118: Traffic Equilibrium Methods, Proceedings 1974. Edited by M. A. Florian. XXIII, 432 pages. 1976.

Vol. 119: Inflation in Small Countries. Proceedings 1974. Edited by H. Frisch. VI, 356 pages. 1976.

Vol. 120: G. Hasenkamp, Specification and Estimation of Multiple-Output Production Functions. VII, 151 pages. 1976.

Vol. 121: J. W. Cohen, On Regenerative Processes in Queueing Theory. IX, 93 pages. 1976.

Vol. 122: M. S. Bazaraa, and C. M. Shetty,Foundations of Optimization VI. 193 pages. 1976

Vol. 123: Multiple Criteria Decision Making. Kyoto 1975. Edited by M. Zeleny. XXVII, 345 pages. 1976.

Vol. 124: M. J. Todd. The Computation of Fixed Points and Applications. VII, 129 pages. 1976.

Vol. 125: Karl C. Mosler. Optimale Transportnetze. Zur Bestimmung ihres kostengünstigsten Standorts bei gegebener Nachfrage. VI, 142 Seiten. 1976.

Vol. 126: Energy, Regional Science and Public Policy. Energy and Environment I. Proceedings 1975. Edited by M. Chatterji and P. Van Rompuy. VIII, 316 pages. 1976.

Vol. 127: Environment, Regional Science and Interregional Modeling. Energy and Environment II. Proceedings 1975. Edited by M. Chatterji and P. Van Rompuy. IX, 211 pages. 1976.